# THE WORLD OF
# NANO-BIOMECHANICS

# THE WORLD OF
# NANO-BIOMECHANICS

## Mechanical Imaging and Measurement by Atomic Force Microscopy

by

ATSUSHI IKAI
Tokyo Institute of Technology
Graduate School of Bioscience and Biotechnology
Yokohama, Japan

With Contributions of
R. AFRIN, S. KASAS, T. GMUR and G. DIETLER

ELSEVIER

AMSTERDAM • BOSTON • HEIDELBERG • LONDON
NEW YORK • OXFORD • PARIS • SAN DIEGO
SAN FRANCISCO • SINGAPORE • SYDNEY • TOKYO

Elsevier
Radarweg 29, PO Box 211, 1000 AE Amsterdam, The Netherlands
Linacre House, Jordan Hill, Oxford OX2 8DP, UK

First edition 2008

**Library of Congress Cataloging-in-Publication Data**
A catalog record for this book is available from the Library of Congress

**British Library Cataloguing in Publication Data**
A catalogue record for this book is available from the British Library

ISBN: 978-0-444-52777-6

For information on all Elsevier publications
visit our website at books.elsevier.com

Printed and bound in Italy
08 09 10 11 12   10 9 8 7 6 5 4 3 2 1

*QH*
*513*
*.I33*
*2008*

*B2365474*

# CONTENTS

# CONTRIBUTORS

**Rehana Afrin**

Laboratory of Biodynamics, Department of Life Science, Graduate School of Bioscience and Biotechnology, Tokyo Institute of Technology, 4259 Nagatsuta, Midori-ku, Yokohama 226-8501, Japan

**S. Kasas**

Laboratoire de physique de la matière vivante, Ecole Polytechnique Fédérale de Lausanne, 1015 Lausanne, Switzerland
Département de Biologie Cellulaire et de Morphologie, Université de Lausanne, Bugnion 9, 1005 Lausanne, Switzerland

**T. Gmur**

Laboratoire de mécanique appliquée et d'analyse de fiabilité, Ecole Polytechnique Fédérale de Lausanne, 1015 Lausanne, Switzerland

**G. Dietler**

Laboratoire de physique de la matière vivante, Ecole Polytechnique Fédérale de Lausanne, 1015 Lausanne, Switzerland

# PREFACE

Nano-biomechanics, the title field of this book, is currently emerging as a new and attractive area of scientific research bridging biological and mechanical sciences at the molecular level. Biomechanics without the prefix of nano has been a quite active field dealing mainly with macroscopic bodily movements and especially with the dynamics of blood flow. In nano-biomechanics, a variety of newly developed devices with the capability of observing and manipulating individual atoms and molecules are ambitiously applied to elucidate the principles of life supporting molecular interactions. I myself am not a physicist or mechanist but a biochemist working in this exciting field and am interested in the material nature of bio-molecules and bio-structures that triggered the emergence of life some four billions of years ago, and since then, have been supporting proliferation of life so successfully.

Since I started working with atomic force microscopes almost 20 years ago, however, I have experienced some problems in bridging biochemistry and mechanics of materials in my own work. I realized that using equations that related the measured quantities to the mechanical parameters of the material for the interpretation of experimentally obtained data was one thing but it was quite another matter to understand the background of those equations. In an applied field such as this, proved equations are picked up from various different sources of mechanics, presenting difficulty for a new comer to find right textbooks every time he or she encounters new equations. This book is meant to be of some help in such occasions and deals exclusively with the proven results of classical mechanics currently used in the measurement of material properties of proteins and cells at the single molecular and single cellular levels.

Thanks to a recent instrumental development that is nurturing an enormous enthusiasm among scientists and engineers to create a new field of nanotechnology, some of the traditional barriers that existed between biological and physical sciences are now rapidly disappearing, at least, at the molecular level. It is naturally true that ultimate goals of physical scientists and those of biological scientists are different, but all of us have shared interests in the behavior of molecules, small or large, and in a new possibility of manipulating them by directly touching each one of them.

Since biological macromolecules are not electrically conductive, biological information transfer is performed mechanically, not electronically as in the case of computer technology, through direct contacts of participating atoms and molecules. The activity of an enzyme, for example, is commonly modulated through binding and unbinding of effector molecules to the enzyme. At the cellular level, a ligand molecule as a carrier of extracellular physiological information binds to a membrane associated receptor triggering a relay of mechanical information transfer from the outside to the inside of the cell.

Examples such as given above are abound in biology that prompt us to consider mechanics as an important and indispensable tool in understanding the basics of biology and developing a new engineering methodology for hand-manipulating proteins, DNA and cells. One of the diverse purposes of manipulation is in the development of new bio-medical technologies.

Using mechanics as a manipulation tool at the molecular level requires us to understand at least the essence of the mechanics of materials which has a long and outstanding history in physics and engineering. For many of us with background in biology, molecular biology, biochemistry or chemistry, the level of undergraduate education in mechanics is rather limited and an extra effort is required to understand the working principles of instruments for the measurement of mechanical properties of materials and to interpret the results of such measurements. Most of the research papers in the relevant fields to nano-biomechanics are written on the assumption that the readers are familiar with elementary mechanics as well as with the background of derivations of many of the equations vital to the interpretation of data. It is time consuming at least, though not impossible, to find references to the required knowledge from a vast array of textbooks on mechanics and to understand the background of the final equations to be used in the measurement and interpretation of data.

This book is essentially a collection of basic equations in macroscopic continuum mechanics that are necessary to understand research papers in biomechanics at the nano-meter and nano-newton level. I tried to explain how such equations were derived from the basic principles of linear mechanics, hoping that this book will save time for those who are coming into this new field and looking for a concise compilation of necessary knowledge from various disciplines of classical mechanics. The subject of this book is mainly static mechanics and, as a result, such otherwise important subjects in nano-biomechanics as viscoelasticity, fluid dynamics and non-linear mechanics, for example, are not treated or only briefly introduced. Readers are recommended to consult with popular textbooks in appropriate fields. Some readers may find this book too elementary or filled with too many equations because I tried to fill in the background derivations, even elementary, as much as possible so that those who have hitherto not been familiar with mechanics can see the meaning of equations and enjoy the process of developing mechanical way of thinking.

As stated above, since the emphasis of the book is in the exposition of basic mechanics, examples of application work are not at all exhaustive and not meant to be. My apology is due here to many authors of important work which I could not refer to in this book. Examples are rather taken to illuminate basic ideas of applying mechanical principles to the study of biological macromolecules and structures built upon them, and many of them are from the work done in the Laboratory of Biodynamics of Tokyo Institute of Technology where I work. My special thanks go to the publishers and individual authors who generously granted me the right to reproduce cited figures from their publications.

I would like to thank many friends and colleagues who supported me in writing this book; among them special thanks are due to Dr. R. Afrin for her contribution of the section of Case Study on carbonic anhydrase II in Chapter 8 and Drs. S. Kasas, T. Gmur and G. Dietler who kindly contributed Chapter 12 on the application of the finite element method. My special thanks are extended to Drs. H. Sekiguchi and I. Harada for preparing some of the figures, Dr. M. Miyata for providing an original photograph of mycoplasma, Drs. R. Afrin, A. Yersin and H. Sekiguchi for proof reading of the original manuscript and Mr. A. Itoh for preparing the cover design. I am, however, solely responsible to the mistakes and inappropriate explanations which may be found in the book, and expect to receive kindly comments from a wide spectrum of readers through e-mail (ikai.a.aa@m.titech.ac.jp).

I also like to express my sincere gratitude to Drs. O. Nishikawa, S. Morita, and M. Tsukada, among many others, for introducing me to the field of nano-mechanics of atoms and molecules and encouraging me to continue the work in the field. I also like to thank my past and present colleagues and graduate students with whom I had and am having excellent opportunities to work together elucidating the exquisite natures of bio-macromolecules and biological structures.

Finally but not in the least, I like to thank Ms. Kristi Green, Ms. Donna de Weerd-Wilson, Mr. Ezhilvijayan Balakrishnan and Mr. Erik Oosterwijk at Elsevier who were concerned in the production of this book and of great help to me.

Atsushi Ikai
September, 2007

CHAPTER ONE

# FORCE IN BIOLOGY

## Contents

## 1.1 WHAT ARE WE MADE OF?

The main theme we are going to explore in this book is a question of 'What kind of materials are we made of'? Our body is soft and fragile compared with many inanimate objects in this world, man-made or not. Can't we have a rock-hard body so that, in a car accident, the car is the one that is crashed and we are the ones to survive? If it is not possible, since our body is said to be the result of self-assembly of a large number of molecules, can we, in the future, control the assembly and disassembly processes of our molecules so that we may, at least, reassemble them after injuries dysfunctionalized our body? Artificial manipulation of atoms, molecules, cells, and tissues of our body is essentially the subject of 'Nano-biomechanics'. For the manipulation of such bodily objects in a distant future, we need to know the physical properties of the materials that make up our body. Since our body functions more like a mechanical device rather than an electronic computer, we investigate the mechanical properties of the bodily components, namely, proteins, nucleic acids, polysaccharides, lipid assemblies, biomembranes, cells and so on by using the state-of-the-art technologies available to us at the present stage. It is especially important to realize that the most abundant bodily components, proteins, are electrically nonconductive, and therefore the information transfer within and between protein-based structures is mainly conducted through their mechanical deformations.

Since mechanical manipulation is performed with an application of force to the sample objects, we will explore, in the first chapter, the meaning of force from our daily experience. Force is something that can be felt and is a more familiar concept than thermodynamic functions such as enthalpy or entropy. This book deals with the effect of force on very small scale because we will be talking about atoms and molecules, and eventually about living cells that are still less than 1 mm in size. Atoms are very strongly built of protons, neutrons, and electrons and will not break down in our body, except for a tiny fraction of radio isotopes. Molecules are clusters of atoms bonded together by covalent bonds, which are also quite strong and difficult to break, but are much weaker compared with the force operating at the nuclear level. Molecules can be converted from one form to another by creating, breaking, and/or exchanging covalent bonds, often, with the help of a catalyst. Catalysts, when used in industry, convert nitrogen gas into ammonia in one notable example, and in the living organisms tens of thousands of them are at work, converting food stuff into the parts of our body and into energy to be consumed for our daily activity.

Catalysts in our body are called enzymes. One of them, called invertase, binds a sugar molecule, for example, and converts it into glucose and fructose by breaking the covalent bond connecting them in the original sugar molecule. Binding a specific substrate molecule from among millions of similarly looking molecules is the most important first step for any enzyme. Binding in this case is promoted by weaker forces associated with 'noncovalent interactions' or 'noncovalent bonds'. Our bodily movements and actions are the result of these noncovalent interactions between and among tens of thousands of molecules in our body. In the next few chapters, we will investigate types of interactions operating at the molecular level in living organisms. To live means to perform daily activities and, to do so and to improve the performance, organisms have developed many exotic devices built of proteins, nucleic acids, lipids, and carbohydrates as the major source materials. We will investigate the basic physical nature of the thus-created devices and the source materials at the molecular level.

## 1.2 HUMAN BODY AND FORCE

### 1.2.1 Gravity and hydrodynamic force

We feel force by the sensation to the muscles of our body when we lift a weight against the gravity, or when we suddenly accelerate our car, for example. Since force is the product of mass $(m)$ and acceleration $(a)$, we feel it when the car is accelerating, but do not feel it when the vehicle is gliding at a constant speed (*i.e.*, when $a = 0$). When we ride a roller coaster, we feel gravitational as well as centrifugal force as the coasting car swishes around sharp corners. We also feel force as a shock when someone suddenly pushes us from behind. A sudden application of a force as described in the above case is called an 'impact'. Force is, thus, something that can be felt as a bodily sensation when we are pushed around.

We feel the gravity of the earth because we, with a big and heavy body, live in air which has much less density than our body. If you live in water like whales and fishes do, you probably don't feel the gravity that much because the gravitational force is largely compensated by the flotation effect in water. As you go down the scale to much smaller levels, you encounter the life without the sensation of gravitational pull. Bacteria, for example, swim around up and down, left and right without feeling much of the gravitational pull. They feel a much more strong effect of viscosity of water. As the scale of your body becomes small, viscosity effect becomes predominant compared with the inertial effect of mass. The Reynolds number $(R_y)$ gives a rough estimate of the relative effects of inertial force versus viscous force.

$$R_y = \frac{\rho R v}{\eta} \tag{1.1}$$

where $\rho, R, v,$ and $\eta$ are the density, the characteristic size and the velocity of the moving body, and the viscosity coefficient of water, respectively. If the Reynolds number is approximately less than 2000, the flow pattern around the moving body is smooth without any turbulence and called a laminar flow, whereas when $R_y > 2000$, the flow tends to be turbulent. In both cases of flow, the moving body experiences inertial resistance as it pushes a body of water aside and viscous resistance from the water sticking to the entire body surface. In a turbulent flow, the moving body must experience a drag force due to an eddy current. The force acting on a spherical object of radius $r$ in a laminar flow is given by the Stokes' law mentioned below, where $f, \eta,$ and $v$ are, respectively, frictional coefficient, fluid viscosity, and velocity of the sphere, of which 2/3 comes from viscosity of water and 1/3 from the pressure effect of water.

$$F = f \times v = 6\pi\eta r v \tag{1.2}$$

For a bacterium of approximate diameter of 1 $\mu$m and swimming in water at a speed of 1 $\mu$m per second, the force is approximately 0.02 pN, which is quite small compared with the force generated by the flagella system of the bacterium. In a micro- to nanometer-scale world, viscous force predominates over the inertial resistance or the eddy current drag, and the magnitude of the viscous dragging can be calculated according to the Stokes' law. The frictional coefficient must be changed depending on the shape of the moving body. Frictional coefficient for nonspherical bodies can be approximated by the method developed by Garcia [1, 2] or by fitting to analytical expressions for prolate or oblate ellipsoids [3, 4]. According to Ref. [1], an approximate value of the frictional coefficient of an object of any shape can be obtained by modeling the shape of the object by an assembly of spheres of radius $R_i$ with the center-to-center distance $r_{ij}$ from another sphere of radius $R_j$ by using the following equation. The method was applied by Ikai to obtain the frictional coefficient of complex proteins [5].

$$f = \frac{6\pi\eta \sum\limits_{i=1}^{N} R_i}{1 + \dfrac{2 \sum\limits_{i=1}^{N} R_i \sum\limits_{i=1}^{N} R_i \frac{R_i R_j}{r_{ij}}}{\sum\limits_{i=1}^{N} R_i R_i}} \qquad (1.3)$$

Hydrodynamic force acting on micro–organisms and molecules is quite small, however, it is an important factor for understanding their behavior.

The effect of various types of force acting on macroscopic biological structures has been studied in the field of biomechanics. Sophisticated analysis based on mathematical formulation of mechanics is applied to understand the response of the biological structures toward externally applied forces. Comprehensive treatises are found in the literature [6, 7].

## 1.2.2 Frictional coefficients

The frictional coefficients of a cylinder of length $L$ and radius $r$ in laminar flow are given in Table 1.1. Here, the subscripts denote whether the flow and the cylinder are parallel or perpendicular to each other, respectively [3].

**Table 1.1**   Frictional coefficients of various bodies in laminar flow.

| Frictional direction | Cylinder $(L \gg r)$ | Ellipsoid $(b \gg a)$ | Sphere |
|---|---|---|---|
| $f_{\text{parallel}}$ | $\dfrac{2\pi\eta L}{\ln(L/2r) - 0.20}$ | $\dfrac{4\pi\eta b}{\ln(2b/a) - 0.5}$ | $6\pi\eta r$ |
| $f_{\text{perpendicular}}$ | $\dfrac{4\pi\eta L}{\ln(L/2r) + 0.84}$ | $\dfrac{8\pi\eta b}{\ln(2b/a) + 0.5}$ | $6\pi\eta r$ |
| $f_{\text{rotational}}$ | $\dfrac{\frac{1}{3}\pi\eta L^3}{\ln(L/2r) - 0.66}$ | $\dfrac{\frac{8}{3}\pi\eta b^3}{\ln(2b/a) - 0.5}$ | $8\pi\eta r^3$ |
| $f_{\text{axial rotation}}$ | $4\pi\eta r^2 L$ | $\frac{16}{3}\pi\eta a^2 b$ | $8\pi\eta r^3$ |

$\eta, L, r, b$, and $a$ are, the viscosity coefficient of the fluid, the length and the cross-sectional radius of cylinders, and the long and short axes of ellipsoidal bodies, respectively. For frictional coefficients for a cylinder near a plane surface, see Ref. [3]. Reproduced from [3] with permission.

## 1.3 BIOMECHANICS AS THE BIG BROTHER

The discipline of biomechanics itself has a long history. This branch of science focuses on the mechanical principle of the function and the movement of our body, and thus deals mainly with macroscopic mechanics. The basis of the discipline is the highly developed theoretical and experimental mechanics with a long history of brilliant work and with highly useful applications to the construction and materials in industry, to name only a few. Though it is a branch of engineering mechanics, much work has also been done in relation to a broad spectrum of medicinal branches including sports and rehabilitation medicine, and people with a variety of backgrounds are working in the discipline. There is a deep scientific and industrial commitment to biomechanics and the World Congress of Biomechanics is a large meeting (held in Munich, Germany, in 2006). A variety of topics were discussed in the congress including the mechanics of muscle contraction, blood flow, organ development, effect of injuries, artificial limbs, sports medicine, cellular mechanics and, as a recent addition, the molecular-and cellular-level mechanics in our body. Nano-biomechanics can be considered an offspring of biomechanics in the sense that it deals with the effect of force on biomolecules and biostructures of the order of nanometer and nanonewton ranges. The principle of mechanics is the same in biomechanics and nano-biomechanics, but the methods to measure a small force and its effect on biosystems are different from those used in macroscopic biomechanics. Classical mechanics deals mainly with materials that are homogeneous in composition and large in their scale compared with the size of a test probe. A comprehensive treatment of biomechanics is given by Fung [6].

Recent development of various physical methods to measure small forces and small displacements has encouraged researchers interested in biomacromolecules and cellular structures to elucidate the relationship between the magnitude of applied force versus that of the deformation occurring in their samples (stress–strain relationship) at the molecular level. By establishing such relations experimentally and by applying theoretical predictions, we can extract mechanical parameters inherent to the material properties of the sample. Thanks to the technological and theoretical advancement in nanoscience and nanotechnology fields, it is now possible to push and/or pull a single molecule of proteins to obtain the force to unfold it from a compact globule to a linearly extended string. The resulting response curves tell us about the rigidity and tensile strength of the intramolecular structure. A similar experiment is now also possible on a single strand of DNA, the very genetic material. The method has been applied to elucidate the mechanistic principle of folding DNA of a total length approximately 1 m into the cellular nucleus of an approximate diameter of a few micrometers.

In this book, the material properties of biological macromolecules and structures self-assembled from the former are expounded and, in Table 1.2, such parameters are presented, with brief notes for each of them.

**Table 1.2**  List of mechanical parameters used in this book.

| Parameters | Notations | Resistance against |
|---|---|---|
| Young's modulus (modulus of elasticity) | $Y$, $(E)$ | Elongation and compression |
| Modulus of rigidity | $G$, $(\mu)$ | Distortion |
| Poisson's ratio | $\nu$, $(\sigma)$ | Thinning ratio accompanying elongation |
| Bulk compressibility | $\kappa$ | Compression under isotropic pressure |
| Torsional rigidity | $\tau$ | Twisting |
| Flexural rigidity | $\kappa$ | Bending |

Note: Notations in parentheses are commonly used alternative expressions.

 ## 1.4 MOLECULAR BASIS FOR STRUCTURAL DESIGN

The basic principle of the structural design of biological systems is building everything bottom up 'from molecules', meaning that macroscopic components of the body are all built directly from molecules and molecular interactions. Since there are no construction workers around, our body is built on the principle of self-assembly of constituent molecules. In contrast, the crane at the construction site, as an example, is made of a relatively small number of macroscopic members of explicit designs, each member being rigid and unbending. They are joined by a team of assembly men or robots using a relatively small number of flexible joints that will maintain the freedom of motion of each member, and its motion is controlled by the force delivered from a centrally located electric motor(s) through a network of wires. The human arm execute a similar task as a crane in a smaller scale, but the motion of an arm is controlled directly by a collection of myocytes (muscle cells) and the force is generated directly from the molecular motion of protein filaments in the cell. Thus, the biological systems are built on the dynamic interactions of a large number of molecules, mainly proteins. Proteins are linear polymers of 20 kinds of amino acids and are tightly folded into specific 3D conformations that are programmed most appropriately for their unique functions.

On the individual basis, proteins function as enzymes, antibodies, receptors, channels, inhibitors, and hormones, and in organized forms, they function as microtubules, muscle filaments, tendons, bones, teeth, hair, silk fibers, to name a few. Thousands of enzymes are known, for example, each catalyzing a different reaction in a concerted way, so that thousands of biochemical reactions proceed in a controlled fashion to keep the host organism happily alive. As stated above, an enzyme has the capacity to bind only one kind of molecule, called substrate, out of millions of others to its active site and carry out a necessary transformation to the molecule. Binding of a substrate to the active site, thus, is the first step of the enzyme catalysis. Such binding is basically a mechanical process in the sense that the substrate is attracted and being directed to the active site through mechanically guided processes. And when finally the substrate is closing in on to the enzyme's active site, the 3D conformation of the enzyme is altered to accommodate the substrate into its active site, forcing the former to change into a different, somewhat distorted conformation from its most stable state, especially around the bond(s) to be broken. The distorted conformation is similar to the activated state of the substrate in the reaction pathway leading to the product. The bound substrate thus sits 'activated' in the active site of the enzyme. This activation is done without raising the temperature but at the expense of the energy of binding to the active site. The active site of the enzyme has strategically located functional side chains of amino acid residues to hold the substrate in a geometry resembling its activated state, thus facilitating the conversion from the reactant to the product under ambient conditions. In the activated state, the substrate is supposed to be in a mechanically strained conformation and the enzyme must be rigid enough to sustain the strain for an extended period in which the reaction proceeds with a higher probability.

Binding, to be precise, specific binding is an important central issue in biochemistry. Many proteins work in association with other molecules, endlessly repeating mutual binding and unbinding processes. When binding molecules are small, they are called ligands and the protein is called receptor, but very often ligands are specific parts of macromolecules such as DNA, proteins, and polysaccharides. In such cases, the macromolecules are also called ligands. Antibodies are a good example of binding proteins without catalytic functions. They constitute a closely related family of proteins with a common 3D structure but each having a different affinity to a selected ligand molecule called an antigen. Antigen binding to an antibody is similar to substrate binding to an enzyme, but antibodies neither activate their ligands nor catalyze their transformation to other molecules. An attempt to modify the arrangement of amino acid residues in the binding site of an antibody so that it would activate the bound ligand was successfully performed by Lerner [8] and the antibody-turned-enzyme is called catalytic antibody or 'abzyme'. A recent review on the catalytic antibodies is found in Ref. [9].

Referring back to an antibody, it rather retains antigen in the stable ground state and forms 3D network gels through its divalent and sometimes multivalent binding capacity. Formation of such a 3D network gel helps clearance of foreign antigens from the host by activating the immune clearance system involving scavenger cells such as macrophages. Antigen binding and the formation of a 3D network by antibody–antigen complexes are largely governed by thermodynamics of the system, but the mechanical stability of the system is also an important factor for the survival of the network in blood.

Another prominent group of binding proteins is receptor proteins, many of which are anchored to the cell membrane through association of their highly hydrophobic, membrane-traversing segments with the hydrocarbon layer of the phospholipid membrane. These proteins are called intrinsic membrane proteins and usually have extracellular and intracellular segments on the opposite ends of the membrane-traversing segment(s). Ligand binding takes place at the binding site on the extracellular segment in most of the cases but occasionally on the intracellular side as well. Ligands bind to a particular receptor to transmit physiological information from other parts of the body, which require particular types of cells to change their biochemical activities in accordance with a new metabolic status of the body. Binding of ligands to the extracellular binding site of the receptor should, therefore, be transmitted to the intracellular metabolic system by way of a conformational change or the aggregation status of receptors.

 ## 1.5 SOFT VERSUS HARD MATERIALS

The functional structures of biological world are realized by the self-assembling property of a large number of the same or different types of molecules rather than by collecting macroscopic members made from bulk solid materials such as metals and/or crystals. Atoms in metals and crystals are linked to each other by strong metallic or covalent bonds, whereas bonding between biomolecules is usually through much weaker types of bonding called noncovalent bonds. They are often not called 'bonds' but rather referred to as noncovalent 'interactions'. It is true that atoms that constitute molecules are held together by strong covalent bonds, but the number of atoms within a macromolecule is a few tens of thousands usually and, at most, a few millions, whereas the number of atoms required to build a single fingertip is in the range of $10^{20}$, which is tens of billions more than the number of atoms in a molecule; therefore, the number of possible interaction mode would be stupendous. Our body is made from a huge array of weak interactions between and among tiny constituents called biomolecules.

The soft and flexible nature of our body enables us to assume millions of subtly different facial features and bodily postures, which are still very difficult to reproduce in robots. Robots are made of metals, plastics, and amorphous materials, all of which are bulk solid materials. The subtleness of the bodily movement of robots is defined by the size of constituent components and the strength of the connecting bonds between them. Since the components of the robotic body are of the order of cm and the strength of the connection between them is of the order of a few newtons; their behavior has a 'centimeter-order' smoothness but not a 'micrometer-order' one, which makes human bodily movements look almost seamless, especially when young and trained, as displayed by top players in Olympic games.

The disadvantage of our soft and flexible body is its weakness revealed on an impact with solid bodies such as rock or a car, as already mentioned above. When involved in a car accident, it is the human body, almost all the time, that is destroyed. To circumvent this disadvantage and to make our body more resistant against violent crashes with something rigid, our ancestors created suits of armor made of small bits of metal, leather, strong fibers, etc. in almost all parts of the world, though their designs were different from place to place. They protected human bodies to a certain extent but with regrettable compromises with some of the graceful bodily functions.

Thus, we understand that our body is not made of solid arms and wires as we see in the design of cranes at a construction site. Muscle is made of billions of billions of protein molecules called myosin and actin, which are bundled into very thin, but still molecular size fibers. The force generating interactions between myosin and actin have been shown to act at a magnitude between zero and a few piconewtons, that is $10^{-12}$ N. A force of 1 N is the weight of 100 ml of water placed on our hand. A piconewton is one trillionth of 1 N, meaning that we need at least $10^{12}$ pairs of myosin versus actin interactions to lift 100 ml of water. By changing the number of myosin versus actin interaction pairs, we can control the force almost continuously.

Whereas the soft parts of biological structures are usually made of proteins, lipids, and carbohydrates, the hard parts such as bones, shells, or scales are mainly composites of inorganic and organic materials. Mammalian bones are made of calcium phosphate complexed with the protein called collagen, and shells of oysters and other shellfish are made of calcium carbonate complexed with the protein called conchiorin. The crustacean shells are composed of the rigid polysaccharide called chitin, complexed with calcium ions. The building principles of such biological hard materials are similar to bulk materials such as concrete used in the construction of man-made buildings.

The most prominent characteristics of living organisms is found, however, in the soft parts such as muscle, heart, liver, skin, and brain. Since the most conspicuous ingredients of the soft parts of the body are proteins, let us take a brief look at them from the viewpoint of materials science. Protein is most familiar to us as meat. Meat is the soft part of the body of animals, mammals among others, and has a texture very similar to that of our body. Proteins are encased inside the cell, which is basically a spherical bag made of phospholipid bilayer membranes as schematically shown in Figure 1.1.

The phospholipid bilayer is approximately 5 nm in thickness and is permeable to water and many nonpolar molecules but impermeable to ionized molecules. The small bag made of phospholipid bilayers has the ability of generating a heterogeneous environment

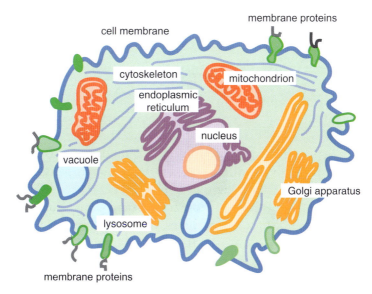

**Figure 1.1** A schematic view of the cross-section of an animal cell. Only major cytoplasmic organelles are shown.

in an otherwise homogeneous medium. Within the encasement of phospholipid bilayers, the most abundant macromolecules are proteins of thousands of different kinds and functions. The characteristics of each protein is determined by the number of amino acid residues covalently linked to one polymeric molecule and by the order of their alignment from one end to the other end of the molecule. A particular alignment of the amino acid residues is called the amino acid sequence, and it generates a special biological function(s). Thus, a polymer having a particular sequence is recognized as a new protein and is given a special name. All amino acids have an amino group and a carboxy group attached to the central $\alpha$-carbon atom as shown in Figure 1.2.

The remaining valencies of the $\alpha$-carbon atom are occupied by a hydrogen atom and the 'side chain' group. Those amino acids that have the amino group on $\alpha$-carbon atom are called $\alpha$-amino acids. The chiral nature of the $\alpha$-carbon is fixed in L-form. The difference in the side-chain structure classifies amino acids into 20 groups, but the rest of the structure around the central $\alpha$-carbon atom is common to all the 20 kinds. Theoretically, the list of $\alpha$-amino acids is endless, but the living organisms utilize only 20 to 22 kinds of them to build proteins in the cellular machinery.

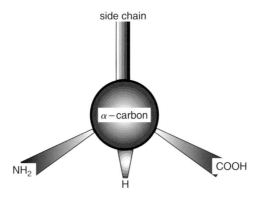

**Figure 1.2** Structure of L-$\alpha$-amino acid. There are 20 (recently expanded to 22) kinds of side chains that form the basis of life-supporting characteristics of tens of thousands of different kinds of proteins.

Proteins are synthesized in the cell as a linear polymeric chain whose amino acid sequence is determined by the genetic code stored in the genomic DNA. The main function of the genome is to store the amino acid sequence of tens of thousands of different kinds of proteins and, occasionally, to allow the cell to read the stored information to produce copies of proteins needed at particular moments in the cell cycle. The stored information is read out in the form of mRNA inside the nucleus and exported to the cytoplasm. The protein synthesis machinery in the cytoplasm is called the ribosome, which is made of a small and a large subunit, each of which is again composed of several kinds of RNAs and proteins. Ribosomes have a function of combining amino acids one after another at the expense of high-energy phosphate bonds according to the instruction contained in the mRNA. At the end of ribosomal polymerization step, proteins are born in the form of a linear chain of hundreds and thousands of amino acid residues, which is called a polypeptide having no biological functions. What endows nascent polypeptides with biological functions is the process called 'folding.'

After biosynthesis, the polypeptide is given a few seconds of free time to search for a thermodynamically most stable conformation under given physiological conditions, which usually leads the polypeptide to a well-defined, compact conformation called the native state. Christian Anfinsen is credited to have shown unequivocally that the thermodynamically most stable conformation of a given polypeptide happens to be the native state of the protein having a specific biological function [10, 11]. He extracted the protein called bovine pancreatic ribonuclease A in a fully functional form and disrupted its native conformation and biological activity by adding urea in high concentration, which was known to destroy most of the noncovalent segmental interactions. He also destroyed disulfide cross-links in the native molecules by reducing them with a reagent called 2-mercaptoethanol. After making sure that the protein was not functional any more, he dialyzed out urea and 2-mercaptoethanol and showed that a substantial fraction of the denatured protein resumed the original biological activity and the characteristic conformation of the native state. The recovery of the reduced disulfide bonds to their unique original combinations out of 105 different possibilities was credited as the most important accomplishment.

Thus, it is a common knowledge today that a protein molecule spontaneously takes on a native conformation, provided it is given the freedom to do so. Such freedom is sometimes or often compromised inside the cell due to the crowded condition therein. In such cases, a certain class of proteins generically called chaperones come in to rescue, providing coveted freedom for each individual polypeptide to fold in a segregated condition [12–15]. Basically chaperones catch an unfolded polypeptide and sequester it inside of its own cavity, and keep the latter for a few seconds until it successfully folds into a correct native conformation.

## 1.6 BIOLOGICAL AND BIOMIMETIC STRUCTURAL MATERIALS

Humans show strong desire to mimic subtle and efficient biological functions by using synthetic materials. One prominent reason for this desire is to improve the fragile nature of biological functions to more durable ones so that they can be used in a convenient manner. There is an effort to create soft materials such as proteins and protein-based structures from synthetic polymers and their composites with inorganic materials and use them as either temporary or permanent replacements for injured or damaged tissues in a patient. To choose appropriate synthetic materials as best fit for the human body, *i.e.*, the one with the best biocompatibility, we have to know the physical properties of the original tissues in addition to their chemical and biochemical properties. It is a common practice to measure mechanical constants of the candidate materials considered relevant for the replacement of hard tissues such as bones and teeth. It is also important to measure or estimate the mechanical constants of muscles and blood vessels that would surround the replacements. Improvements of the surface properties of biomimetic structures are one of the most intensely explored field together with the recent advancement in the production of biodegradable polymers and ceramics.

## 1.7 WEAR AND TEAR OF BIOLOGICAL STRUCTURES

Can proteins and other biological structures repeat catalytic or binding activities endlessly without wear and tear? When a stress is applied to a macroscopic object which is perfectly elastic, its shape changes and strain energy accumulates inside, but when the stress is removed, the shape returns to the original one and the strain energy returns to zero. In practical cases, certain rearrangements of the atomic-level structure take place while the material is in a strained state, and the eventual removal of stress does not let all the atoms go back to the original positions; thus the material does not regain its original shape. Such is the case of partially plastic materials. If this process is to be repeated tens of thousands of times, continuous rearrangements of atoms and bond breaking will eventually form a small cavity or a crack within the material, which will lead to the fracture of the material.

A general theory and experimental observations on the fatigue of materials is summarized in Ref. [16]. Propagation of preexisting cracks was conceived as an important factor of mechanical breakdown of materials by Griffith in the 1920s [17]. He conjectured from the material constants of materials with a microscopic crack that the tensile strength of materials would be much less than expected because an externally applied force (or in terms of stress that is defined in Chapter 2 as force per unit cross-sectional area) tends to concentrate around the edge of the crack and a brittle fracture of material takes place as a result of crack propagation (Figure 1.3). Fatigue is a special kind of fracture that occurs after application of external force for an extended duration of time and is most often observed in the fracture of metallic objects.

The critical stress $\sigma_c$ required for crack propagation in a brittle material is given by

$$\sigma_c = \left( \frac{2Y\gamma_s}{\pi a} \right)^{1/2} \tag{1.4}$$

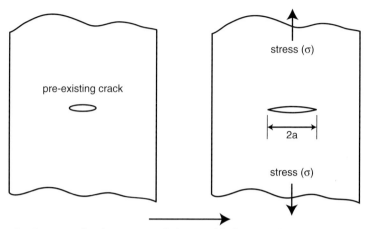

Crack propagation due to an applied stress (s). Curvature becomes zero at the edge of the crack.

**Figure 1.3** Crack propagation in an imperfect 2D material. A preexisting crack in (a) is deformed by the application of longitudinal stress so that the curvature at both ends of the crack becomes zero. Stress concentration takes place at the two edges resulting in the propagation of the crack along horizontal directions as in (b).

where $Y$ is Young's modulus, $\gamma_s$ specific surface energy, and $a$ half of the length of an internal crack.

Equation (1.4) was derived by considering a balance between the change in the elastic energy around the crack $(-\pi a^2 \sigma^2 / Y)$ and the increase in the surface energy $(4a r_s)$ due to the crack extension.

The biological nanostructures constantly function for the production and transport of indispensable metabolites of life. Functioning of biostructures is usually accompanied by the conformational changes of small and large scales, which may be the source of wear and tear of such structures if we take analogies from our daily macroscopic examples. For example, the engine of a car requires periodic attention and care to keep it running. Biological nanostructures are surprisingly durable against wear and tear. A good example is the red blood cell which circulates in our body approximately 200,000 times during its life-time of approximately 120 days [18]. During its circulation through blood vessels, it must endure tortuous journeys of passing through capillaries only half as wide as its own diameter of 8 $\mu$m. Inside the capillaries it changes its form from the usual biconcave one to a bullet type. After red blood cells repeat this shape change for hundreds of thousands of times, they retain almost all their biological activities, but they are constantly replaced with new ones with an average half-life of 120 days.

If the surface of an object is constantly in the abrasive contact with another surface, a certain fraction of surface atoms will be constantly torn away, with the process leading to the wearing of the surface. Our bodily surface is covered with skin, *i.e.*, the epithelial cells, which are constantly being renewed from the inside after shedding of the old ones. This constant replacement of old bodily components with the newly formed ones is the basic principle of coping with wear and tear of constituent materials in biological world. You can see it in plant life as well when trees shed leaves after using them from spring to autumn for photosynthesis and respiration and wait for the new ones to bud in spring, renewing their synthetic life.

A question arises then, do biomolecules, or more in general terms, do molecules wear and tear? Under the influence of light, especially that of UV light, x-rays, or strong electric field, molecules are decomposed, but under mechanical strain, molecules are not broken down easily. In conclusion, the structural changes of biological molecules and structures responsible to their functions are basically repeatable millions and billions times and there seems to be little wear and tear of them.

## 1.8 THERMODYNAMICS AND MECHANICS IN NANOMETER SCALE BIOLOGY

Life processes are often mentioned not in terms of a thermodynamic equilibrium but rather in terms of nonequilibrium, dynamics states. This statement is basically correct but, in most of the biochemical experiments, we regard a section of life processes presented in front of us as in a thermodynamic equilibrium. However, in every day life, we regard many natural and artificial phenomena to be in equilibrium; from the universal point of view, the earth itself is in nonequilibrium state including all the phenomena taking place on its surface. For example, as stated previously, the native state of an enzyme has been shown to be thermodynamically in the most stable state for a given covalent structure, *i.e.*, the primary structure. Many other seemingly dynamic features of biological processes are, when dissected down to constituent molecular interaction levels, based on equilibrium thermodynamics of molecular events. Thus, when locally viewed, life is based on equilibrium relationships between and among molecules. Nonequilibrium nature of life is, however, manifested when we think of the total bodily metabolism where there are daily uptake and disposal of food and waste materials. The most important aspect of being an open and nonequilibrium state is in the capability of constantly shedding old materials and replacing them with newly acquired or synthesized ones. In a sense, there is an endless supply of raw materials that can be converted into useful components of life processes together with efficient ways to get rid of what is not needed any more. Spurious proteins are degraded by the ubiquitin–proteasome system, programmed death occurs in unwanted cells, etc. It is most important to emphasize that organisms extract chemical energy from the food stuff and use that energy to convert other raw materials to useful molecules to sustain life.

## Bibliography

[1]  Bloomfield, V., Dalton, W. O., and van Holde, K. E. (1967), Frictional coefficients of multi-subunit structures, 1. Theory, *Biopolymers*, 5, 135–148.

[2]  Garcia de la Torre, J. and Bloomfield, V. A. (1977), Hydrodynamic properties of macromolecular complexes, 1. Translation, *Biopolymers*, 16, 1747–1763.

[3]  Howard, J. (2001), 'Mechanics of Motor Proteins and the Cytoskeleton', Sinaur Associates. Sunderland, MA.

[4]  van Holde, K. E., Johnson, C., and Ho, P. S. (1998), 'Principles of Physical Biochemistry' (2nd ed.), Prentice-Hall. Englewood Cliffs, NJ.

[5] Ikai A. (1986), Calculation and experimental verification of the frictional ratio of hagfish proteinase inhibitor, *J. Ultrastruct. Mol. Struct. Res.*, 96, 146–150.

[6] Fung, Y. C. (1993), 'Biomechanics: Mechanical Properties of Living Tissues', Springer. New York, NY.

[7] Winter, D. A. (2004), 'Biomechanics and Motor Control of Human Movement', Wiley. New York, NY.

[8] Lerner, R. A., Benkovic, S. J., and Schultz, P. G., (1991), At the crossroads of chemistry and immunology: catalytic antibodies, *Science*, 252, 659–667.

[9] Keinan, E. (ed.) (2005), 'Catalytic Antibodies', Wiley VCH. New York, NY.

[10] Anfinsen, C. B. and Haber, E. (1961), Studies on the reduction and re-formation of protein disulfide bonds. *J. Biol. Chem.*, 236, 1361–1363.

[11] Haber, E. and Anfinsen, C. B. (1961), Regeneration of enzyme activity by air oxidation of reduced subtilisin-modified ribonuclease, *J. Biol. Chem.*, 236, 422–424.

[12] Ellis, R. J. (1987), Proteins as molecular chaperones, *Nature*, 328, 378–379.

[13] Gething, M. J. and Sambrook, J. (1992), Protein folding in the cell, *Nature*, 355, 33–45.

[14] Horwich, A. L., Neupert, W., and Hartl, F. U. (1990), Protein-catalysed protein folding, *Trends Biotechnol.*, 8, 126–131.

[15] Hartl, F. U. (1996), Molecular chaperones in cellular protein folding, *Nature*, 381, 571–580.

[16] Suresh, S. (1998) 'Fatigue of Materials' (Cambridge Solid State Science Series) (2nd ed.). Cambridge University Press. Cambridge, UK.

[17] Griffith, A. A. (1921), Phenomena of rupture and flow in solids, *Phil. Trans. Roy. Soc.*, 221, 163–198.

[18] Yawata, Y. (2003), 'Cell Membrane: The Red Blood Cell as a Model', Wiley VCH. New York, NY.

## CHAPTER TWO

# INTRODUCTION TO BASIC MECHANICS

## Contents

## 2.1 ELASTIC AND PLASTIC DEFORMATION OF MATERIALS

Structural samples (often called members in engineering terminology) undergo deformations when a tensile, compressive, or shear force is applied. The tensile force elongates, compressive force shortens, and shear force distorts them. Upon removal of the applied force, the shape of an elastic sample returns to the original one, and, if not, the sample is partially or totally plastic. In many cases, samples show an elastic deformation while the applied force is small and as the force becomes greater, contribution of plastic deformation gradually increases.

During an elastic deformation, the covalent and noncovalent bonds that hold the atoms in the sample body are not broken but displaced from their equilibrium distance, *i.e.*, deformed due to the application of an external force, but on removal of the force all the deformed bonds return to their equilibrium positions, recovering the original shape of the object. The proportionality constant of the stress–strain relationship in the elastic regime of deformation is called Young's modulus. The general introduction to the mechanics of materials is given by Timoshenko [1] and an advanced version is presented in Ref. [2]. A concise treatise of the theory of mechanics is found in Ref. [3]. The latter is especially suited to understand the general validity of linearized treatment of the elastic material.

If the bonds that keep the sample in a fixed shape are weak or if the applied force is large, consequently under a large deformation regime, some of the covalent or noncovalent bonds are broken and therefore cannot restore the original equilibrium positions. As a result, the shape of the sample material is permanently changed, and such a deformation is termed plastic deformation. Unless all the bonds are broken, the sample shape is partially restored. The deformation of an ordinary material sample often proceeds as shown in Figure 2.2 given later in this chapter. The range of strain of the elastic deformation regime is small for materials such as steel or silicon but large for rubber or plastics.

## 2.2 STRESS AND STRAIN RELATIONSHIP

When the material properties of sample specimens should be compared, a certain level of normalization of applied force and that of resultant deformation is necessary. The same magnitude of a tensile force applied to two specimens made of the same material but having different diameters produces elongations whose magnitudes

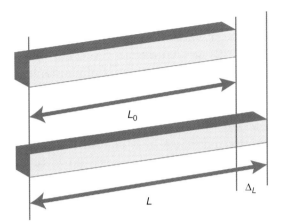

**Figure 2.1** Elongation of a rectangular member under a tensile force along the axial direction. The relative elongation, $\Delta L/L_0$, is defined as strain ($\epsilon$) and the force per unit cross-sectional area, $F/A$, is defined as stress ($\sigma$). In linear mechanics, they are linearly related as $\sigma = Y\epsilon$, where $Y$ is a constant called Young's modulus.

are inversely proportional to the cross-section of the sample on the two faces of force application. Therefore, when we are interested in the material constant rather than in the absolute magnitude of the elongation, the force ($F$) must be divided by the cross-sectional area $A$ and normalized as 'stress' ($\sigma = F/A$). Similarly, if one of the specimens is twice as long as the other, both made of the same material, the same stress causes twice as large elongation on the former as on the latter. Here again, the elongation ($\Delta L$) should be normalized with respect to the original length of the specimen ($L_0$); therefore, 'strain' ($\varepsilon = \Delta L/L_0$) is defined as the elongation per unit length. For a relatively small strain, the stress follows the Hooke's law and is proportional to the strain, and the proportionality constant is defined as Young's modulus $Y$.

$$\sigma = Y\varepsilon \qquad (2.1)$$

For the basic tenets of linear mechanics, an excerpt from Ref. [3] is given in the Appendix for reference. It is a clear exposition of the basic mechanics.

When the strain is small, the above relationship applies to both a positive and a negative strain; therefore, elongation ($\varepsilon > 0$) and compression ($\varepsilon < 0$) are symmetric. Stress has the dimension of N (newton) per m$^2$, whereas strain is dimensionless. Therefore, $Y$ has the dimension of N/m$^2$, which is given a special name, pascal, abbreviated as Pa. Obviously for a rigid material such as steel, $Y$ is as large as $1-3 \times 10^{11}$ Pa = 100–300 GPa, and for soft materials such as plastics and wood, $Y$ is in a smaller range in the order of $10^9$ = 1 GPa and for rubber it is even down to $10^6$ = 1 MPa. Roughly, man-made objects have Young's modulus between 1 and 1000 GPa, whereas the bodily materials of living organisms have young's modulus within 1/1000 of such values, $i.e.$, between 1 and 1000 MPa (see Table 2.1).

The plastic nature of the materials comes from their liquid-like property. If the material is totally plastic, the sample deformation never stops and the sample is liquid-like. The resistance of the sample against the applied stress is only temporary due to the internal viscosity of the liquid.

**Table 2.1**  Young's moduli of biological materials.

| Material | Young's modulus (GPa) |
|---|---|
| **Proteins** | Actin 2, Tubulin 2, Coiled coil 2, IF protein 2, Flagellin 1, Silk 5, Lysozyme 5, Carbonic anhydrase 0.08, Denatured protein 0.002, Abductin 0.004, Resilin 0.002, Elastin 0.002 |
| **Sugars** | Cellulose 20–40, Chitin 45 |
| **Nucleic acid** | DNA 1 |
| **Composites** | Teeth 75, Shell 68, Bone 19, Wood 16, Muscle 0.040, Cartlage 0.015 |
| **Others** | Rubber 0.001, Plastics 2, Concrete 24, Glass 71, Steel 215 |

Note: The young's moduli are determined using the method described in Ref. [6] with some modifications.

## 2.3 MECHANICAL BREAKDOWN OF MATERIALS

When a tensile force is continuously applied to a sample body, initially elongates in proportion to the magnitude of the force (Hookean regime), and after reaching a peak, the force is reduced, with a further elongation of the sample. In this plastic elongation regime, the sample suffers a deformation called 'necking', where a sudden thinning of the sample is observed. The peak force observed before the start of necking is called yield force. After yielding to the tensile force, the sample is further elongated with a small increase of force. After sustaining a maximum force (tensile strength), the sample is finally torn into smaller parts. A series of events in the breaking process is illustrated in Figure 2.2. The tensile force and the corresponding elongation are normalized as stress and strain.

The Hookean regime for hard materials is very narrow on the abscissa in Figure 2.2, and a sharp and linear increase of stress results in a straight line with a large slope corresponding to a high value of Young's modulus. Whereas soft and elastic materials such as rubber show a wide range of Hookean regime. Necking is a prominent feature in plastics and metals. The breaking behavior of biological materials at the molecular level is not well known yet and will be the subject of nano-biomechanics in the near future.

 ## 2.4 VISCOELASTICITY

Materials that exhibit a pronounced contribution of fluid-like viscosity in addition to elasticity are called viscoelastic materials. Their mechanical properties depend on the time scale of force application. If a force is applied over a short duration of time, their behavior is almost solid-like, but if the force is applied slowly, they behave like fluids. Although the molecular nature of viscoelastic materials is complicated and may differ from one material

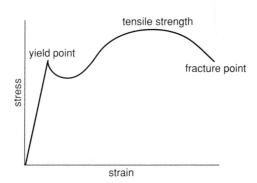

**Figure 2.2**   The breakdown process of a sample under application of tensile stress, $\sigma$, on the ordinate, and the corresponding strain, $\epsilon$, on the abscissa.

Note: Deformation and ultimate breakdown of mechanical members generally proceed along the line in this figure. The initial linear region is called Hookean regime where the stain is reversible, which is followed by a sudden breakdown of the linear relationship and a large deformation with a small increase of stress. The point of maximum stress is called the tensile strength, which is soon followed by the final breakdown. The region of nonlinear extension of the material is possible with a supply of material to the extending region from other regions and is made possible through the mechanism called 'necking.'

to another, their viscoelastic behavior has a common factor and is modeled by a few typical types by using three mechanical elements: (1) a mass for inertia, (2) a spring for reproducing the elastic property, and (3) a dashpot for reproducing fluid property. Models differ based on how the three elements are connected, as illustrated in Figure 2.3.

- *Maxwell model*: a spring is in series with a dashpot representing a viscoelastic fluid.

- *Voigt model*: a spring is in parallel with a dashpot representing a viscoelastic solid.

- *Complex model*: Features of both Maxwell and Voigt models are incorporated.

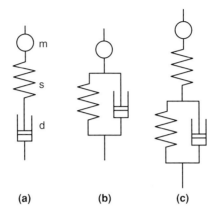

(a)                 (b)                 (c)

**Figure 2.3**   Two models of viscoelastic materials. (a) Maxwell model: a mass, m; a spring, s; and a dashpot, d are conneced in series to represent liquid-type materials. (b) Voigt model: the two elements (s) and (d) are connected in parallel to represent solid materials. (c) model. When $m = 0$, the vertical displacement, $z(t)$, for an instantaneous application of a force $F$ at $t = 0$ is $z(t) = (F/\eta)t + (F/k)$ for (a), $z(t) = (F/k)[1 - \exp(-kt/\eta)]$ for (b) and $z(t) = (F/k_1) + (F/k_2)[1 - \exp(-k_2t/\eta)]$ for (c), where $k$'s and $\eta$ are, respectively, the force constants of the spring(s) and the viscosity coefficient and the dashpot. Corresponding differential equations for (a) and (b) are, respectively, $dz/dt = (F/\eta) + (dF/dt)/k$ and $dz/dt = -(k/\eta)(z - F/k)$.

## 2.5 MECHANICAL MODULI OF BIOLOGICAL MATERIALS

### 2.5.1 Mechanical deformations

How rigid are biological materials, such as proteins, DNA, viruses, cells, and many others, in terms of Young's modulus and how can we measure their rigidity? This question is the major theme of this book. In general, biological samples are as soft as or much softer than synthetic plastics. In the literature, several different methods to estimate the Young's modulus of various proteins have been described.

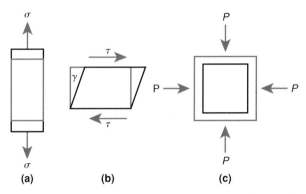

**Figure 2.4** Deformation of materials can be classified into three types: (a) tensile deformation, (b) shear deformation, and (c) triaxial (isobaric) deformation.

To begin with, we will look at three different types of deformations as shown in Figure 2.4.

Since definition of Young's modulus in tensile deformation is already given, the remaining two types of deformation and the mechanical modulus associated with them will be explained below.

### 2.5.2 Shear deformation and rigidity modulus

A shear deformation is caused by an application of a pair of force along two opposing faces of a rectangular body so that there is a change in the angle $\gamma$ as in Figure 2.4b. Within the approximation of linear mechanics, the magnitude of $\gamma$ is proportional to the applied force per unit area $(F/A : \text{shearstress}(\tau))$ with a proportionality constant of $G$, the rigidity modulus. $G$ is often expressed as $\mu$ as well.

$$\frac{F}{A} = G\gamma \tag{2.2}$$

### 2.5.3 Triaxial deformation and bulk compressibility

When the surface of a rectangular body of volume $V_0$ is under stress along three normal directions, deformation is given by the change in its volume, $\Delta V$. In the case where all three stresses (pressures) are of the same magnitude, it is called a spherical stress. In this case, $\Delta V/V_0$ is proportional to the pressure $P = F/A$ and the bulk modulus of elasticity $K$ and inversely proportional to the bulk compressibility, $\kappa$. If $\kappa$ is large, the volume change under the same pressure is large, and vice versa.

$$P = K\frac{\Delta V}{V_0} = \frac{1}{\kappa}\frac{\Delta V}{V_0} \tag{2.3}$$

### 2.5.4 $Y$, $G$, and $K$ are all related through Poisson's ratio

Let us first look at the relationship between Young's modulus $(Y)$, rigidity modulus $(G)$, and bulk compressibility $(\kappa)$ as given below, where $\nu$ is Poisson's ratio, which will be explained later.

$$Y = 2G(1 + \nu) = 3(1 - 2\nu)/\kappa \tag{2.4}$$

The relationship between $Y$ and $G$ can be derived based on a simplified case as shown in Figure 2.5.

Suppose a rectangular body (a) is distorted to rhomboid (b). The distance between $b$ and $d$, $L_{bd}$, is elongated to

$$L_{bd} = \sqrt{2}h(1 + \varepsilon_{\max}) \tag{2.5}$$

where $\varepsilon_{\max}$ is the strain along the line $bd$ due to a normal stress on the face $ac$. Remember that $\gamma$ is small and $ab = ad = h$ still after

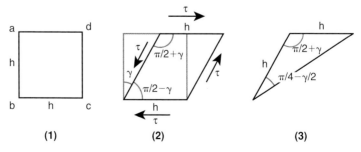

(1)                    (2)                    (3)

**Figure 2.5** A diagram explaining realtionship between Young's modulus, $Y$, and rigidity modulus, $G$. Parameters appearing in the text are shown in the figure.

deformation. $L_{bd}$ can be expressed in a different way by applying a trigonometric relation.

$$L_{bd}^2 = h^2 + h^2 - 2h^2 \cos\left(\frac{\pi}{2} + \gamma\right) \tag{2.6}$$

By equating the two expressions for $L_{bd}$, we obtain

$$(1 + \varepsilon_{max})^2 = 1 - \cos\left(\frac{\pi}{2} + \gamma\right) \tag{2.7}$$

Therefore,

$$1 + 2\varepsilon_{max} + \varepsilon_{max}^2 = 1 + \sin\gamma \tag{2.8}$$

Assuming that $\varepsilon_{max}^2$ and $\gamma$ are small, the former is set to 0 and $\sin\gamma$ is expanded as $\gamma$. Thus,

$$\varepsilon_{max} = \frac{\gamma}{2} \tag{2.9}$$

By definition, $\gamma = \tau/G$ and $\varepsilon_{max} = \tau(1+\nu)/Y$, where $\tau$ is the tangential force acting on the sides of the rectangle. According Chapter 3 of Gere and Timoshenko [1], the latter equation comes from the fact that each pair of $\tau$'s produces a normal stress $\sigma_{max} = (2 \times \tau/\sqrt{2})/\sqrt{2} = \tau$ on the face $ac$ (45° against $cd$). The area along $ac$ is $\sqrt{2}$ times larger than faces $ab$ or $cd$. $\sigma_{max}(=\tau)$ elongates $L_{bd}$ by $Y\varepsilon_{max}$ and a normal compressive stress generated on $ac$ which is equal to $-\tau$ elongates $L_{bd}$ by $\tau/\nu Y$. The combined effect of shear force $\tau$ on $L_{bd}$ is $\varepsilon_{max} = \tau/Y + \tau\nu/Y = \tau(1+\nu)/Y$. Therefore, we obtain

$$G = \frac{Y}{2(1+\nu)} \tag{2.10}$$

Next, we will look at the case of triaxial stress where the strains produced by the stresses, $\sigma_x$, $\sigma_y$, and $\sigma_z$ acting independently, are summed to obtain the resultant strains as below.

$$\varepsilon_x = \frac{\sigma_x}{Y} - \frac{\nu}{Y}(\sigma_y + \sigma_z) \tag{2.11}$$

$$\varepsilon_y = \frac{\sigma_y}{Y} - \frac{\nu}{Y}(\sigma_z + \sigma_x) \tag{2.12}$$

$$\varepsilon_z = \frac{\sigma_z}{Y} - \frac{\nu}{Y}(\sigma_x + \sigma_y) \tag{2.13}$$

These equations can be solved for the stresses in terms of strains and the result is

$$\sigma_x = \frac{Y}{(1+\nu)(1-2\nu)}[(1-\nu)\varepsilon_x + \nu(\varepsilon_y + \varepsilon_z)] \tag{2.14}$$

$$\sigma_y = \frac{Y}{(1+\nu)(1-2\nu)}[(1-\nu)\varepsilon_y + \nu(\varepsilon_z + \varepsilon_x)] \tag{2.15}$$

$$\sigma_z = \frac{Y}{(1+\nu)(1-2\nu)}[(1-\nu)\varepsilon_z + \nu(\varepsilon_x + \varepsilon_y)] \tag{2.16}$$

When the three stresses have the same value of $\sigma_0$ for homogeneous and isotropic body, the three strains also have the same value of $\varepsilon_0$

$$\varepsilon_0 = \frac{\sigma_0}{Y}(1 - 2\nu) \tag{2.17}$$

Then, the unit volume change $e$ which is equivalent to $\Delta V/V_0$ is given as

$$e = 3\varepsilon_0 = \frac{3\sigma_0(1-2\nu)}{Y} = \frac{\sigma_0}{K},$$
$$\text{where } K = \frac{Y}{3(1-2\nu)} = \frac{1}{\kappa} \tag{2.18}$$

In summary,

$$Y = 2G(1+\nu) = 3(1-2\nu)K = \frac{3(1-2\nu)}{\kappa} \tag{2.19}$$

Material having a small bulk compressibility, $\kappa$, is rigid and its Young's modulus and rigidity modulus are large so that it is difficult to stretch or deform it. To understand the rigidity of a protein molecule, $Y$ or $G$ must be measured at the single molecular level, but the measurement of bulk compressibility, $\kappa$, can be done on a solution of a protein by measuring the velocity of sound wave, $(c)$, propagating in the solution. The sound velocity and the adiabatic compressibility, $(\kappa)$, is related by the following equation, where $\rho$ is the density of the liquid. In a liquid medium, the adiabatic compressibility is close to the isothermal one and can be replaced with the latter.

$$c = \sqrt{\frac{1}{\kappa\rho}} \tag{2.20}$$

From the measured value of bulk compressibility, one has to separate the contribution of the solute (protein) and that of the solvent. Once the compressibility of the protein is determined, $Y$ may be estimated from Eq. (2.19) but the value of Poisson's ratio is not known for proteins. It is often assumed that the value of $\nu$ is close to that of synthetic polymers, namely, $\nu \sim 0.35$. Since the estimation of $Y$ from $\kappa$ is sensitively dependent on the value of $\nu$ through $(1 - 2\nu)$ factor, especially when $\nu$ is close to 0.5, it is safe to avoid the use of Eq. (2.20) for the estimation of $Y$ unless the value of $\nu$ is known fairly accurately.

### 2.5.5 What is Poisson's ratio?

When a rod is axially placed along the $x$ axis and pulled along $x$ direction with an elongation of $\Delta x$, the width along $y$ and $z$ directions will be reduced by $\Delta y$ and $\Delta z$, respectively. The change in the volume of the sample will be

$$\Delta V = (x_0 + \Delta x)(y_0 + \Delta y)(z_0 + \Delta z) - x_0 y_0 z_0 \tag{2.21}$$
$$= x_0 y_0 \Delta z + y_0 z_0 \Delta x + z_0 x_0 \Delta y \tag{2.22}$$
$$\frac{\Delta V}{V_0} = \frac{\Delta x}{x_0} + \frac{\Delta y}{y_0} + \frac{\Delta z}{z_0} = \varepsilon_x + \varepsilon_y + \varepsilon_z$$
$$= \varepsilon_x(1 - \nu_{xy} - \nu_{xz}) \tag{2.23}$$

$\nu_{xy}$ and $\nu_{xz}$ are Poisson's ratios defined as $\nu_{xy} = -\varepsilon_y/\varepsilon_x$ and $\nu_{xz} = -\varepsilon_z/\varepsilon_x$, respectively.

When there is no volume change, and assuming for a rod of square cross-section where $y_0 = z_0$ and $\Delta y = \Delta z$, we have $\varepsilon_y = \varepsilon_z$, and therefore, $\nu_{xy} = \nu_{xz} = \nu$. If there is no volume change before and after pulling, $\Delta V = 0$, which results in

$$1 - 2\nu = 0, \quad \text{thus } \nu = 0.5 \tag{2.24}$$

For all of $Y, G$, and $\kappa$ to be positive $((1 + \nu) > 0$ and $(1 - 2\nu) > 0)$, it is required that $-1 \leq \nu \leq 0.5$. Usually, $\nu$ lies between 0.2 and 0.5. A material having $\nu = 0.5$ is called incompressible. Vulcanized rubber has a $\nu$ close to 0.5. For materials with a negative value of $\nu$, we expect volume expansion under tensile stress, which is rather unintuitive but has been shown to be conceivable and could be real.

## 2.6 FLUID AND VISCOSITY

The difference between elastic materials and fluid is that in the latter, the interactions among constituent molecules are so weak that the intermolecular bonds are constantly formed and broken within a short time range compared to the case in which external force is applied. When the external force changes its magnitude or direction, no memory of the previous force is retained in the fluid, except for a very short duration of time. Even though the intermolecular bonds in fluid are weak in general, each fluid differs in the average strength of such bonds, which determines its viscous nature together with the size of the fluid molecule. The stronger the intermolecular bonds and the larger the molecular size, the more viscous is the fluid.

Water is the most important fluid for organisms on the earth, and it is more viscous compared with other fluids made of molecules of similar size because the intermolecular bond between water molecules, termed the hydrogen bond, is stronger than other forces responsible for maintaining the fluid states of other molecules. The dominating character of hydrogen bond in water confers it with many peculiarities, which actually help life sustain on the earth. Water freezes and vaporizes at significantly higher temperatures than other liquids of similar molecular size. For example, the boiling temperatures of $H_2S$ and $H_2Se$ are 212 K and 232 K, respectively, compared with 373 K for ordinary water. The solid water is less dense than liquid water at the same temperature; therefore, ice floats in liquid water.

Without water of such physical and chemical nature, the emergence and development of life would be probably impossible on this planet. Our body is said to be composed of 70% water in weight, which means that most of the cells have a similar water content. Water as a bulk solvent and water as a participant in chemical reactions ceaselessly occurring in our body are vital to our existence. So when we think about fluid and fluid viscosity in this book, it means water viscosity. The viscosity coefficient of water is close to 0.0012 Pa·s at 20°C and about twice as large at 0°C.

## 2.7 ADHESION AND FRICTION

If you can see molecules inside the cell, they must be floating around in a rather crowded environment, because thousands of different kinds of proteins and equally heterogeneous populations of small molecules are packed, altogether making up nearly 30% of the weight of the cell. Since each of them must find proper counterparts to have specific interactions at a small force of a few pNs, it is vitally important to exclude the possibility of nonspecific interactions to the maximum extent. Nonspecific interactions between surfaces of biological components are thus kept minimal through the long history of molecular evolution. When such components are taken out of a biological environment and brought to a forced contact with an artificially prepared surface of materials such as glass, silicon, mica, or gold, significant levels of unwanted adhesion takes place. This is a well-known obstacle in the application of biological materials for industrial and medical uses. Biosensors, for example, utilize electron-transfer reactions on a solid electrode surface with specific enzyme molecules, glucose oxidase, in the case of a glucose sensor. In many cases, key enzymes are protected from strong adhesive interactions with the electrode by the presence of inert proteins such as bovine serum albumin. Detailed molecular-level investigations of protein adhesion to solid surfaces are attracting interests of researchers in diverse fields because of its biological and industrial importance. It is an especially important field, where computer simulations can make significant contributions since the details of molecular and submolecular events in adhesion cannot be directly inspected by visual methods.

If adhesion is kept at a minimal level, friction is not important, except in cases where a strong shear force is applied to biological interface. One possible case is again red blood cells traveling through capillary systems. The cells are forced to change their shape from a discoidal biconcave one to a more elongated bullet shape, but the surface of the capillary epithelium and that of the red blood cell are covered with polysaccharides called a glycocalyx layer of a few hundred nanometer in thickness. Both surfaces are negatively charged due to substantial amounts of sialic acids; therefore, they repel each other. Carbohydrate layers are generally very hydrophilic and are hydrated maximally, so that they do not have fixed conformations like protein molecules, and consequently they are viscous but not adhesive. Friction between red blood cells and capillary epithelium is minimized by the presence of glycocalyx layers on both sides.

Friction has been an important engineering issue since a very early stage of human civilization. Frictional force is the force required to move an object sitting on a flat substrate at a constant velocity. To move an object of weight $W$ sitting on a solid surface, first you have to apply a tangential force $F$ from the side of the object. $F$ is known as frictional force and the ratio between $F$ and $W$ is the static frictional coefficient $\mu_s$.

$$\mu = \frac{F}{W} \tag{2.25}$$

Once the object has started moving, less force is needed to keep it moving; therefore, $\mu$ becomes smaller and is called the kinetic frictional coefficient, $\mu_k$, where $\mu_s > \mu_k$. The Amonton–Coulomb law of friction tells us that $\mu_s$ and $\mu_k$ are independent of the contact area between the object and the surface. Moreover, the law also asserts that $\mu_k$ is independent of the moving velocity but has less wide applicability. It is interesting to observe that friction is often higher for the two well-polished interfaces than for unpolished ones.

At the molecular and atomic levels, friction is explained by the formation and destruction of interatomic and intermolecular bonds. Even at the atomic level, the two contacting surfaces are not flat but have convolutions due to the crystalline arrangement of atoms. When the friction between an AFM probe and a flat mica surface was studied, various phenomena were observed, which could be ascribed to interatomic interactions between a small number of atoms. One such observations is called 'stick-slip' friction, meaning that atoms at the very end of the AFM probe repeat sticking to and bouncing off from the atoms on the substrate surface.

The above discussion is mainly concerned with dry friction. In biological measurements, samples are embedded in water and we have to deal with 'wet' friction between polymeric substances as described in relation to red blood cells and blood vessels. At the molecular scale, the intermolecular interaction between polymer chains can be treated as frictional events as observed when a polymer chain is pulled out of entanglement with other polymer chains or from gels or polymer solid. Another interesting area to be explored is the intramolecular friction which is beginning to be considered in the stretching and compression experiments on proteins and DNA.

## 2.8 MECHANICALLY CONTROLLED SYSTEMS

Enzymes are highly efficient catalysts and, therefore, their activities should be kept under strict control according to the physiological conditions of the body. When the body needs a particular metabolite, the enzymes involved in its synthesis are all turned on, but when the body does not need the metabolite any more, the particular metabolic pathway must be shut off. To respond to this requirement of the body rather quickly, the biochemical feedback system is used. If the concentration of the metabolite in demand becomes high initially, it will bind to the enzyme responsible for its production and inhibits its activity. The most efficient way to turn down the production of the end product of a dedicated metabolic pathway is to stop the activity of the key enzyme that catalyzes reactions at the branching point of the metabolic pathway as shown in Figure 2.6. This type of inhibition scheme is called 'end-product' inhibition or 'negative feed back' system. Since accumulation of the end product leads to termination of the activity of the enzyme at the branching point, supplies of all the intermediate metabolites to the downstream enzymes are reduced, thus economizing the cell chemistry. It is good that none of the intermediate metabolites between the branching point and the end product does not accumulate excessively in the cell because some of them could have toxic effects on the cell physiology.

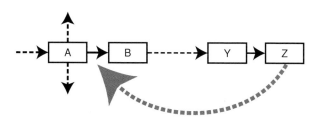

Feedback inhibition or end product inhibition

A: precursor at the branching point

B----Y: intermediates

Z: end product

**Figure 2.6** An example of biological feedback systems is represented in the form of an idealized cascade metabolic system where the end product of the final enzyme reaction inhibits the first enzyme after the branching point.

Now, we look at the molecular details of end-product inhibition. We first note that the chemical structure of the end product is very different from that of the natural substrate for the enzyme at the branching point; thus the former cannot bind to the active site of the enzyme and functions as an efficient competitive inhibitor there. Instead, the end product finds a more comfortable binding site on the surface of the enzyme away from its active site and, after binding, indirectly influences the activity of the enzyme. It has been currently considered that binding of an effector to the enzyme surface forces the latter to change the conformation of its active site in sequential steps and eventually modulates the enzyme activity by slight distortion of its active-site geometry. In thermodynamic terms, the enzyme is assumed to have two conformationally different states: one active without effector binding and the other inactive with a bound effector. In the absence of the effector, the population of the latter is very small, but in the presence of the effector, the equilibrium is shifted to the inactive form.

If we try to control the activity of such enzymes without using effectors, we may do so by changing the conformation of the enzyme by applying force. Addition and removal of effector is time-consuming and consequently the time required for the modulation of enzyme activity may be excessively long for industrial use of the controlled enzyme activity. For such a purpose, development of a solid-state enzyme device is necessary whose activity can be switched on and off within a very short time, say in milliseconds, by an application of either tensile or compressive force. Such possibility of at least modulating protein activity has been tried both experimentally and theoretically [4, 5]. By applying compressive force on green fluorescence protein (GFP) molecules bound to a solid surface by a colloidal AFM probe, it has been shown that the fluorescent activity of GFP is partially suppressed with a mechanical distortion. The molecular dynamics simulation of this process revealed that the chromophore of GFP is forced to undergo a rotation around a particular single bond, changing the dihedral angle to the direction of low fluorescence.

# Bibliography

[1] Timoshenko, S. P. and Gere, J. M. (2002), 'Mechanics of Materials', PWS Publishing Co. Boston, MA.

[2] Timoshenko, S. P. and Goodier, J. N. (1970), 'Theory of Elasticity' (3rd ed.), McGraw Hill. Aukland, NewZealand.

[3] Landau, L. D. and Lifshitz, E. M. (1986), 'Theory of Elasticity' (3rd ed. English), Butterworth-Heinemann. Oxford, UK.

[4] Kodama, T., Ohtani, H., Arakawa, H., and Ikai, A. (2005), Mechanical perturbation-induced fluorescence change of green fluorescent protein. *Appl. Phys. Lett.*, 86, 043901-1–043901-3.

[5] Gao, Q., Tagami, K., Fujihira, M., and Tsukada, M. (2006), Quenching mechanism of mechanically compressed green fluorescent protein studied by CASSCF/AM1, *Jpn. J. Appl. Phys.*, 45, L929–L931.

[6] Howard, J. (2001), 'Mechanics of Motor Proteins and the Cytoskeleton', Sinaur Associates. Sunderland, MA.

## CHAPTER THREE

# FORCE AND FORCE MEASUREMENT APPARATUSES

## Contents

## 3.1 MECHANICAL, THERMAL, AND CHEMICAL FORCES

Force causes the deformation of material bodies and force can be applied in various different ways and force is force no matter what its origin is, just like energy is energy no matter how it is generated. But in practice, we distinguish different forms of energy such as heat (thermal energy), light (electromagnetic energy), and electrostatic. Similarly, by knowing different origins of force, we may have an access to control particular types of force.

Force $(F)$ is the negative derivative of potential $(V)$ with respect to the distance $(r)$ of movement and expressed in various different ways as below.

$$F(r) = -\frac{dV}{dr} \tag{3.1}$$

When the distance is given by the three components, $x$, $y$, and $z$,

$$F(x,y,z) = -\left( \frac{\partial V(x,y,z)}{\partial x}, \frac{\partial V(x,y,z)}{\partial y}, \frac{\partial V(x,y,z)}{\partial z} \right) \tag{3.2}$$

$$= -\left( \frac{\partial}{\partial x}, \frac{\partial}{\partial y}, \frac{\partial}{\partial z} \right) V(x,y,z) \tag{3.3}$$

$$= -\operatorname{grad} V = -\nabla V \quad \text{where } \nabla \text{ is called}$$

the gradient or nabla operator. $\tag{3.4}$

Force is a vector and has a magnitude and a direction and is derived from a scalar function $V$. Force played the central role in the Newtonian mechanics, but gradually energy established a more central role in physics, especially in quantum mechanics.

Measuring the magnitude of force acting on experimental samples and the mechanical response from the latter are the major concerns of this section. If we can measure the magnitude and direction of force acting on atoms and molecules, then we will be able to control the force in the known range to influence the life processes as we observe under microscopes. Measuring a small force acting at the molecular and atomic levels has been a difficult undertaking because the magnitude of the force involved is in the order of $10^{-9}$ times smaller compared with the force we are experiencing daily, i.e., the force of several newtons (N). Recently, there has been an explosive development in the technology of measuring even smaller forces of $10^{-12}$ N. Today, scientists are talking about measuring femto newton forces, i.e., $10^{-15}$ N. Let us take a look at some of the recently developed instruments with a capability of measuring such small forces acting in the invisible world of atoms and molecules, the nano-world.

## 3.2 LASER TRAP

One of the most sensitive force-measuring devices is called laser trap or laser tweezers consisting of a focused laser beam and a microscope. The laser tweezers technique has been developed by Ashkin [1] and has been greatly improved and widely used since then. It utilizes the force exerted by the light at the interface of two materials with different refractive indices.

It has been shown that a metallic or plastic particle whose diameter is less than the wavelength of light can be trapped by a focused laser beam. When a converging laser beam is irradiated on the particle, the scattering force, $F_{scat}$, and the gradient force, $F_{grad}$, work on the particle in two different ways. $F_{scat}$ always pushes the particle in the direction of light propagation; thus a particle between the beam source and the focus is pushed to the focal area, but one on the other side of the focal area is pushed away. Whereas the gradient force, $F_{grad}$, drives the particle to move in the direction of increasing electromagnetic field; therefore, it is called the gradient force, and it pulls particles to the focal area from all directions. The gradient force, acting on a particle of polarizability $\alpha$ in a medium of refractive index $n_1$, is given in the following form [2].

$$F_{grad} = \frac{1}{4} n_1 \alpha \nabla \left( |E|^2 \right) \qquad (3.5)$$

$E$ is the electric field in the laser beam. For a spherical particle, the polarizability $\alpha$ is given by the following form.

$$\alpha = 4\pi \varepsilon_r \varepsilon_0 \frac{n_r^2 - 1}{n_r^2 + 2} r^3 \qquad (3.6)$$

where $n_r$ is the refractive index of the particle relative to that of the surrounding medium and $r$ is the radius of the particle. The gradient force is thus dependent on the volume of the sample particle through the $r^3$ term. In general, $\alpha$ is a complex number, but when its real part is positive, the particle is attracted to the stronger light field according to the gradient of the force. For example, a small gold particle with $r \ll \lambda$ is pulled into the laser-beam spot by the gradient force because the real part of $\alpha$ is positive for gold. The variation of the gradient force and the scattering force in the vicinity of the laser focus is given in Ref. [1].

In Figure 3.1, the laser beam is irradiated from the left to the right region, and its focal point is shown in the middle of the figure. If the latex bead is in the left region from the focal point of the laser beam, the force operating at the interface pushes the bead to the direction of the focal point according to the force given by Eq. (3.6), and if the bead is in the right of the focus, the force acts to push it to the focus. In either way, the bead is pulled to the direction of the focus. Moreover, if the bead is located at the same level as the focus but partially out of focal point in the lateral direction, it is again pushed to the focus from either side of the focus because the electric field becomes stronger toward the focus.

There is always a force to pull the latex bead back to the center of the focus of the laser beam, and the bead is 'trapped' there. Thus, another term for this phenomenon is the 'optical trap' method. Since the intensity of the laser beam has a somewhat Gaussian distribution in the 2D cross-section of the beam, the potential energy and the force associated with it can be calculated.

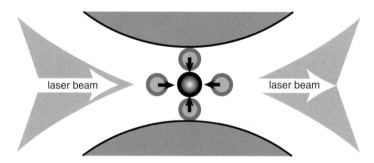

**Figure 3.1** Qualitative view of the trapping of dielectric spheres in the focal point of laser beam. The beam comes in from the left, forms a focal point in the center and leaves to the right. The refraction of a typical laser beam gives scattering and gradient forces whose vector sum is always restored for axial and transverse displacements of the sphere from the focal point.

There are two different ways to calculate the strength of the gradient force according to the relative size of the particle with respect to the wavelength of the laser light $\lambda$. The gradient force is in the 'ray optics' regime when the radius of the particle ($r$) is much larger than the wavelength of the laser beam ($r \gg \lambda$) and in the Rayleigh regime when $r \ll \lambda$. In the former case, the gradient force is independent of the size of the particle and proportional to the gradient of $nP/c$ ($n$: refractive index of the medium, $P$: power of the laser beam, and $c$: the velocity of light), whereas in the latter regime, it changes with $r^3$ because the polarizability is proportional to the volume of the particle as we have seen. In typical experimental cases in biology, $r \sim \lambda$ and the size dependence of the gradient force are not accurately known. The trapping force is then experimentally calibrated by dragging a spherical particle by laser tweezers and calculating the dragging force according to the hydrodynamic frictional force based on the Stokes law of spherical particle, i.e., $F = 6\pi\eta r v$, where $\eta$ and $v$ are the viscosity coefficient of the medium and the constant rate of particle movement, respectively [3]. As an example of force measurement of laser trap, Hénon et al. reported the result of incident laser power versus trapping force on the latex bead of 1.05 $\mu$m as estimated by the hydrodynamic drag method as given in Figure 3.2 [4].

The result in Figure 3.2 shows that the trapping force is almost linearly proportional to the laser power but slightly dependent on the distance from the cover slip. The maximum force available in this case was approximately 80 pN.

The magnitude of the gradient force and that of the scattering force depend on the laser power. If the laser power is high and if the difference in the refractive index of the particle is much higher than that of the medium, scattering force becomes larger than the gradient force and the laser trap fails.

For a particle having only a slightly larger refractive index compared with that of the medium, the gradient force is not strong enough as a trapping device. If that is the case, two laser beams employed in opposing directions and focusing at the same position would compensate the scattering force but reinforce the gradient force.

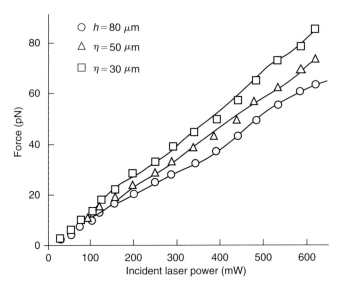

**Figure 3.2** The relationship between the incident laser power on the abscissa and the trapping force in the focus of the laser beam at three different distances from the coverslip as given in the inset as reproduced from Ref. [4] with permission.

This technique can be used to measure the force exerted to the latex bead trapped in the center of the focus by tethering the bead to another system, which mechanically disturbs the position of the bead. For example, suppose the bead is tethered to another bead immobilized at a fixed position through a polymer chain and the laser beam is moved horizontally right and left, the bead trapped in the focus also moves to the right and left, but its position changes slightly from the center of the focus because of the force that acts to extend or compress the polymer chain.

 ## 3.3 Atomic Force Microscope

### 3.3.1 History and principle

In 1982, invention of the scanning tunneling microscope (STM) was announced, and the STM was immediately welcomed by the researchers in surface science and related fields because of its superb capability of showing the arrangement of individual atoms on the metallic and semiconducting surfaces [5, 6]. By measuring the magnitude of the tunneling current between an electrically conductive probe and a sample, which is exponentially dependent on the probe versus sample distance, STM can provide a contour map of the sample surface in terms of the local density of states (LDOS) [7, 8]. Special attention was paid to the arrangement of atoms in the reconstructed silicon (111) surface, where the STM clearly showed a real-space image of the surface, which was very similar to the model that had been presented by Takayanagi et al. [9, 10]. A few years later in 1986, the atomic force microscope (AFM) was invented by Binnig and colleagues [11], which was also capable of imaging atoms and molecules on a solid surface. One advantage of AFM is that it can operate on either conductive, semiconductive, or nonconductive materials, whereas an STM requires electron-conductive material as a sample.

A comprehensive review of the force-mode operation of AFM, and the results reported in this area has been given by Butt et al. [12].

The AFM operates on the principle of mechanical rather than electronical interactions between the probe and the sample surface. Suppose both the probe and the sample are electrically neutral, dielectric materials. As the probe approaches within a few tens of nanometers, it comes into the regime of an attractive van der Waals force (including all of the dipole–dipole, dipole-induced dipole, and London dispersion interactions). The probe is weakly attracted toward the sample surface and as it approaches closer to the sample, the probe is in the repulsive realm in terms of Lennard–Jones potential [13], where the probe is strongly repelled from the sample surface. If one measures either the attractive or repulsive force by spring mechanics that reflects the interaction of the probe with the sample surface, one can reproduce the contour map of the sample surface in terms of sample height.

Both in the original and in the commercial AFMs available presently, the force sensor is a thin cantilever spring of approximately $100\,\mu$m in length $(L)$, 20–30$\,\mu$m in width $(w)$, and less than $1\,\mu\text{m}$ in thickness $(d)$. When it is made of silicon or silicon nitride of Young's modulus $(Y)$ of 100–150 GPa, the spring constant $(k)$ can be in the range of 0.01–10 nN/nm. It may be estimated from the following equation [14].

First, the relationship between the load $F$ and the cantilever deflection at the free end, $y_{\text{max}}$, is expressed as

$$y_{\text{max}} = \frac{L^3}{3YI}F = \frac{4L^3}{Ywt^3}F \tag{3.7}$$

where $I$ is the cross-sectional moment of the cantilever (in the case of cross-section of the rectangle of width $w$ and thickness $t$, $I = wt^3/12$). Therefore, the spring constant is

$$k = \frac{Ywt^3}{4L^3} = \frac{100 \times 10^9 \times 30 \times 10^{-6} \times (1 \times 10^{-6})^3}{4 \times (1 \times 10^{-4})^3} \tag{3.8}$$

$$= 0.75\,\text{N/m} = 0.75\,\text{nN/nm} \tag{3.9}$$

These equations are derived from the consideration of beam-bending problem in mechanics, as detailed in the Appendices.

### 3.3.2 How to use AFM for force measurement

Since AFM is most favored by many researchers for the measurement of pN to nN forces of interaction forces between biological macromolecules, this section is devoted to the use of AFM in the force-measuring mode.

In the operation of AFM, detection of the magnitude of cantilever beam bending (deflection) is most important because the sensitivity of this operation is directly proportional to the value of interaction force between the probe and the sample. To this day, a variety of methods have been tested, but the most popular one involves the use of an optical lever.

• Optical lever method utilizes a focused laser beam irradiated on the back of the cantilever and reflected into the bisected or quadrisected photodiode detector. A small change in the cantilever-deflection changes the incident angle of the laser beam to the back of the cantilever, and consequently the direction of the reflected beam is changed. The difference between the intensity of the light going into the upper and lower halves of the bisected detector gives the cantilever deflection.

- Interference between the two laser beams, one reflected from the back of the cantilever and the reference beam, can be used to detect the magnitude of cantilever deflection.

- Change in capacitance due to the change in the distance between the cantilever and the substrate surface may be accurately monitored by placing an electrode against the back of the cantilever and may be used to record the cantilever deflection.

- Tunneling current between the backside of a conductive cantilever and a sharp metallic probe positioned at a short distance to the back of the cantilever can give an accurate measure of the cantilever deflection. The prototype of AFM was built with this detection system.

- Change in Piezo resistance due to the deflection of a self-actuating cantilever is used in noncontact-type AFM.

In Figure 3.3, a schematic view of the optical lever method of detection of cantilever deflection is given.

When the optical lever method is used, it does not record the magnitude of deflection itself but the slope of the cantilever at the position where the laser beam is irradiated. Since dependency of the deflection on the distance $x$ from the free end is given as

$$y = \frac{FL^3}{YI} \frac{1}{6} \left[ 2 - 3\left(\frac{x}{L}\right) + \left(\frac{x}{L}\right)^3 \right] \qquad (3.10)$$

its slope is

$$\frac{dy}{dx} = \frac{FL^3}{YI} \frac{1}{6} \left[ -\left(\frac{3}{L}\right) + 3\left(\frac{x^2}{L^3}\right) \right] \qquad (3.11)$$

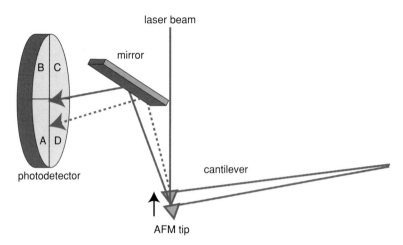

**Figure 3.3**   The principle of the optical lever method used as a force transducer of AFM. The laser beam is reflected from the back of the gold-coated cantilever and reaches the quadrisected (A, B, C, and D) photodiode detector. The differential output (A + D)-(B + C) is proportional to the vertical cantilever deflection from the equilibrium position, and (A + B + C + D) gives the sum value.

which is equal to

$$\frac{-FL^2}{2YI} = \frac{-3}{2L}y_{\max} \quad \text{when } x = 0 \qquad (3.12)$$

Since the tangent of the cantilever at $x = 0$ is proportional to $y_{\max}$ as long as $L$ is constant, *i.e.*, having the laser beam at the same position on the same cantilever, the tangent of the cantilever gives an accurate measure of its deflection.

Another way to detect the magnitude of force being inflicted on the cantilever is to keep it oscillating constantly by supplying vibrational energy at its resonance frequency, keeping its amplitude constant at approximately 1 nm. When the cantilever of force constant $k$ comes into the region of attractive interaction with the sample surface, its oscillatory frequency is slightly diminished because the effective force constant of the cantilever $k'$ is now modulated by the gradient of the force field $F'$, which is negative in this case.

$$k' = k + F' \qquad (3.13)$$

Since the oscillatory frequency of the cantilever is related to the spring constant through the following equation,

$$\nu = \sqrt{\frac{k'}{m}} \qquad (3.14)$$

reduction of $k'$ from $k$ lowers $\nu$ by $\Delta\nu$ from $\nu_0 = \sqrt{k/m}$, which can be accurately measured with a precision better than $10^{-5}$. $\Delta\nu$ is thus related to the force gradient, not force itself, by the following equation,

$$\Delta\nu = \frac{\nu_0}{2k} \frac{\partial F}{\partial z} \qquad (3.15)$$

The above equation was derived on the assumption that the force gradient is constant [15].

This detection method is used in the noncontact AFM (ncAFM), which has been shown to have a true atomic resolution and a capability of atom manipulation on metallic and semiconductor surfaces [16]. Initially, ncAFM was used under high vacuum because, under air or liquid conditions, the cantilever vibration is strongly damped (*i.e.*, vibration under air or liquid conditions is energetically much less efficient compared with that under vacuum) and signal-to-noise ratio is very low. The efficiency of transmitting supplied energy to the cantilever vibration is called Q-factor and is operationally defined as the ratio between the height against half-width of the power spectrum of the cantilever vibration (see Figure 3.4). It is difficult to physically increase Q-factor, but it can be effectively increased by using an electronic control circuit called Q-control.

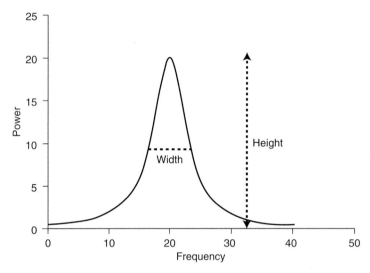

**Figure 3.4** Q-factor represents the sharpness of the resonance peak and thus inversely the degree of energy dissipation in the form of (peak height)/(half width).

## 3.4 BIOMEMBRANE FORCE PROBE

Biomembrane force probe was introduced by Evans to measure the mechanical response of live cells in a culture medium under constant observation by an optical microscope [17]. A live cell is immobilized on the tip of a glass capillary by sucking a part of the cell into the capillary under the application of a negative pressure, and the manipulation is done on the opposite side of the cell, which is freely exposed to the culture medium. In one instance of measuring the interaction force between an intrinsic membrane protein on the cell surface and a chosen ligand or a specific antibody, the ligand or antibody molecules were immobilized on a latex bead. The bead was then bonded to a red blood cell at the mouth of another glass capillary facing head to head to the first capillary as shown in Figure 3.5.

The blood cell is deformed to a spherical shape because of the negative pressure that acts to bold it at the mouth of the capillary.

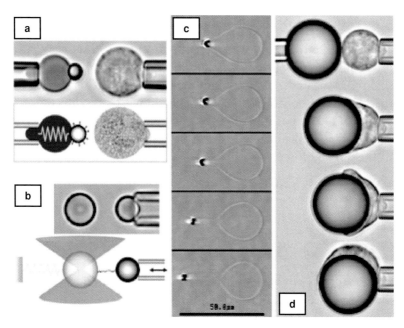

**Figure 3.5**  Examples of nano-to-micro-mechanical experiments. (a) A biomembrane force probe is used to test the adhesion strength between P-selectin and a neutrophil. (b) A combination of optical tweezers and micropipette manipulation is used to study DNA elasticity. (c) A paramagnetic bead is manipulated using a "magnetic puller" as it drags along a lipid vesicle, extruding a membrane tether. (d) A human neutrophil attempts to "eat" a large, antibody-covered bead ("frustrated phagocytosis"). Reproduced with permission of Professor Volkmar Heinrich.

The latex bead is brought into a brief contact with the sample cell on the left and, after a specified time of contact, the right capillary is pulled away from the left one. As it is pulled away, both the sample cell and the red blood cell are deformed due to the tensile force maintained between the two capillaries, *i.e.*, between the ligand and receptor molecules. The magnitude of the tensile force is estimated from the deformation of the spherical red blood cell. The suction force applied at the tip of the pipette can be lower than 0.1 pN.

### 3.4.1 Equation of force transduction

When a force is applied at a single point on the cell surface that is immobilized on the opposite side, the shape of the spherical cell slightly elongated or compressed, and for a small displacement, the changes in axial length have been shown to be directly proportional to the axial force working as a Hookean spring [18]. The axial change in the diameter, $\delta$, has been shown to be proportional to the applied force, $F$, as follows.

$$F = k_f \delta \tag{3.16}$$

$$k_f \sim 2\pi \frac{\sigma}{\ln[4R_0^2/(R_p R_c)]}, \quad \sigma = \frac{1}{2}\frac{R_p}{(1 - R_p/R_0)}\Delta p \tag{3.17}$$

where $R_0$, $R_p$, $R_c$, and $\Delta p$ are the radius of the spherically swollen red blood cell, the radius of the pipette, the radius of the adhesive contact between the glass bead and the cell, and the suction pressure in pipette.

## 3.5 MAGNETIC BEADS

Small magnetic beads internalized into a live cell have been used to probe the local viscoelastic properties of the intracellular fluid by applying a magnetic field from outside the cell [19]. By applying a magnetic field to twist a magnetic bead in the cell, one can measure the rotational frictional coefficient of the bead, and hence the viscosity or viscoelasticity of the surrounding fluid is obtained.

## 3.6 GEL COLUMNS

An interesting recent advance in the measurement of the force exerted by living cells as they slowly move around the surface of

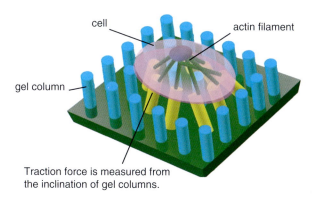

Figure 3.6   A schematic view of the gel column method of force measurement between a moving cell and the column heads. The figure is due to Dr. Ichiro Harada.

the substrate is the use of flexible gel columns [20]. An array of vertical gel columns is prepared on a solid surface by the micro mechanical method, and living cells are placed over a vertical array of the columns. As a cell starts moving, tensile force is created between the adhesive structure of the cell and the upper ends of the columns, and the columns are bent. The degree of bending is measured using an optical microscope and converted to the force under the assumption that the columns are linearly behaving cantilevers. Figure 3.6 gives a schematic view of the experimental setup for this type of experiment.

 ## 3.7 CANTILEVER FORCE SENSORS

The cantilever used in the AFM technology has been shown to be useful in a uniquely different way [21]. An array of cantilevers without the AFM probe part was manufactured and their backside was chemically activated, and single-stranded DNA with a specific nucleotide sequence was end grafted on them. After immersing the modified cantilever, single-stranded DNA with the complementary sequence with that on the cantilever was added to the solution. When hybridization took place between the immobilized DNA and the complementary DNA in solution, the double-stranded DNA tried to occupy a larger space on the cantilever and thus laterally pushed against each other. This lateral expansion of DNA volume caused downward deflection of the cantilevers, which can be easily detected by the optical lever system similar to the one used in commercial AFM instruments. The same principle can be applied to antigen–antibody systems. The idea has been used in the production of commercially available sensor for biological affinity measurement.

##  3.8 LOADING-RATE DEPENDENCE

Once the force to break or distort a particular object is measured, one can explore the dependence of the magnitude of the force on the rate of force loading, or simply loading rate. In daily life, we experience that when tearing off an adhesive tape, a larger force is required to do so rapidly than slowly. In more precise terms, it is not the speed of tearing but that of applying force in terms of force/time, and if the applied force increases linearly as a function of time, it is a constant value for a particular set of experimental parameters.

The functional dependence of the mean force value measured for a particular experimental system depends on the logarithm of the loading rate $(r)$ [22] as schematically shown in Figure 3.7, and it has been experimentally verified.

$$F^* = \frac{k_B T}{\Delta x} \log r + \frac{k_B T}{\Delta x} (\log t_0 - \log F_0) \qquad (3.18)$$

The parameter $\Delta x$ that appears in Eq. (3.18) is shown to be synonymous with the 'activation distance' and is schematically shown in Figure 3.8. It is defined as the elongation of the bond to be broken to its activated state from the equilibrium

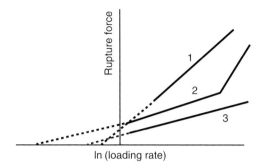

**Figure 3.7** Schematic idea of the dependence of the mean rupture force on the ordinate and of the logarithm of the loading rate on the abscissa. The slope of the linear part of the plot is inversely proportional to the activation distance for unbinding the bond in focus, and the intercept with the $x$ axis gives the unbinding rate constant in the absence of applied force. When the plot has two linear parts of different slopes, the energy diagram for unbinding is interpreted to have two energy barriers.

length. When the bond to be broken is uniquely defined as the bond between two atoms; the meaning of $\Delta x$ is clear but in many reactions involving macromolecular species, the 'bond' has a more conceptual picture. Moreover, when the bond to be broken is not parallel to the direction of the applied force but inclined at an angle $\theta$, $\Delta x$ measured in experiment corresponds to $\Delta x^0 \cos \theta$, where $\Delta x^0$ is the true length of the activation distance.

The slope of the force-loading rate curve is steeper for a smaller value of $\Delta x$ and vice versa. For example, for force rupture of a covalent bond, $\Delta x$ cannot be larger than 0.1 nm, but for rupture of macromolecular bonds such as biotin–avidin bond, $\Delta x$ values as large as 0.5 nm has been reported. In the cases where the force-loading rate graph is divided into two curves with different slopes, the presence of two prominent activation barriers is postulated as shown in Figure 3.8. The application of a tensile force to such a bond system, first, lowers the outer energy barrier with a longer activation distance $\Delta x_2$, and after sufficiently lowering the outer barrier, the presence of the inner barrier with a shorter $\Delta x_1$ is exposed. Thus, the mechanical-rupture experiments

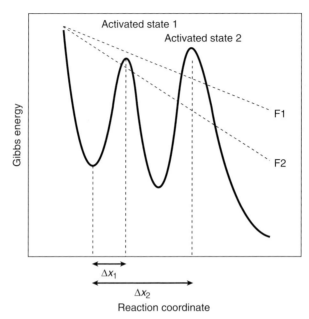

**Figure 3.8**  Schematic diagram of energy of reaction pathway of bond breaking, showing the activation energy and activation distance. In this figure, there are two activation barriers designated as activated state 1 and 2. The applied force is assumed to decrease the activation barrier in proportion to the activation distances times the maginitude of force as shown by two dashed lines.

of an interacting pair of molecules give a schematic view of the energy diagram of the reaction pathway. $\Delta x$ has a similar role as the Arrhenius activation energy to be obtained from the dependence of the reaction rate on temperature. In short, at a lower and a higher loading rate, respectively, the outer and the inner barrier acts as the rate-limiting step.

The rupture force is not a constant of the system because force is not a thermodynamic parameter as entropy, enthalpy, or Gibbs energies are. It makes the comparison of different systems in terms of mechanical strength difficult and, in a sense, meaningless because the rupture force that is larger for one system than another may become lower at different loading rates depending on the slope of the curves.

### 3.8.1 Derivation of the loading-rate dependence of the mean rupture force

Loading rate indicates the rate of increase in the force applied to a target bond to rupture it. It has a dimension of force/time (N/s, nN/s, or pN/s). The magnitude of rupture force measured by the application of a tensile force depends on the logarithm of the loading rate. Here, a brief summary of the deduction of the dependence is presented according to Evans [22]. The theory starts from the exposition of the rate equation for the bond breaking at a given temperature $T$. The rate of bond breaking $k_0$ in the absence of an applied force is given as below, with $E$ being the activation energy for the reaction. $t_0$ is the natural life-time of the bond, and $A$ is a constant.

$$k_0 = A\exp(-E/k_{\mathrm{B}}T) = \frac{1}{t_0} \qquad (3.19)$$

When a tensile force $F$ is applied to pull the target bond, the probability of the bond to pass the activation energy barrier increases by a factor of $\exp[F\Delta x/k_{\mathrm{B}}T]$. Here, $\Delta x$ is the distance from the equilibrium bond length to the activated length, and $F\Delta x$ represents the work done by the applied force and the activation energy barrier is lowered by this amount. This is evidently not the work done to push the reactant to the top of the energy barrier, which would be equal to $E$. $F\Delta x$ is not as large as $E$, but when the bond length occasionally reached the activation distnce, $\Delta x$, even a small work of $F\Delta x$ contributes to help the bond to cross the activation energy barrier by modifying the ordinary Boltzmann factor of $\exp[-E/k_{\mathrm{B}}T]$ to $\exp[-(E - F\Delta x)/k_{\mathrm{B}}T]$, thus increasing the crossing probability by a factor of $\exp[F\Delta x/k_{\mathrm{B}}T]$.

The rate constant of bond breaking under the applied force is

$$k_d = \frac{1}{t_0}\exp\left(\frac{F}{F_\beta}\right), \ \text{ where } F_\beta = \frac{k_{\mathrm{B}}T}{\Delta x} \qquad (3.20)$$

By writing the fraction of unbroken bonds as $S(t)$, the rate of bond breaking is presented as follows. Here, the rate of bond formation is neglected.

$$\frac{\mathrm{d}S(t)}{\mathrm{d}t} = -k_d S(t), \text{ thus } S(t) = S(0)^{-k_d t} \qquad (3.21)$$

Experimental data are usually presented in the form of histogram, where the ordinate is the frequency of observing the rupture force corresponding to the bins in the abscissa. Thus, the histogram represents $F$ versus $\mathrm{d}S/\mathrm{d}F$. The mean of the histogram, $F^*$, is obtained from the condition, $\mathrm{d}^2 S/\mathrm{d}F^2 = 0$.

We introduce 'loading rate $= \mathrm{d}F/\mathrm{d}t$' as an alternative variable of time and rewrite the original differential equation as follows.

$$\frac{\mathrm{d}S}{\mathrm{d}F}\left(\frac{\mathrm{d}F}{\mathrm{d}t}\right) = -k_d S \text{ by setting } \frac{\mathrm{d}F}{\mathrm{d}t} = r \text{ we have } \frac{\mathrm{d}S}{\mathrm{d}F} = \frac{-k_d}{r}S$$
$$(3.22)$$

By replacing the variables in the rate equation and setting $\mathrm{d}^2 S/\mathrm{d}F^2 = 0$, we obtain the value of $F^*$ that satisfies it.

$$\frac{\mathrm{d}^2 S}{\mathrm{d}F^2} = -\frac{1}{r}\left(\frac{\mathrm{d}k_d}{\mathrm{d}F}S + k_d\frac{\mathrm{d}S}{\mathrm{d}F}\right) = 0 \qquad (3.23)$$

where the following identities apply.

$$\frac{\mathrm{d}k_d}{\mathrm{d}F} = \frac{k_d}{F_\beta} \text{ and } \frac{\mathrm{d}S}{\mathrm{d}F} = -\frac{1}{r}k_d S \qquad (3.24)$$

Then, Eq. (3.23) is transformed as

$$\frac{1}{F_\beta}k_d - k_d^2\frac{1}{r} = 0 \qquad (3.25)$$

By dividing both sides by $k_d$, we obtain

$$\frac{1}{F_\beta} - \frac{k_d}{r} = 0 \quad k_d = \frac{r}{F_\beta} \qquad (3.26)$$

which means

$$\frac{1}{t_0}\exp\left(\frac{F^*}{F_\beta}\right) = \frac{r}{F_\beta} \qquad (3.27)$$

By taking the logarithm of both sides, we obtain

$$-\log t_0 + \frac{F^*}{F_\beta} = \log r - \log F_\beta \qquad (3.28)$$

After rearrangement, we finally obtain the loading-rate dependence of the mean force of the histogram as below which is equivalent to Eq. (3.18).

$$F^* = F_\beta \log r + F_\beta(\log t_0 - \log F_0) \qquad (3.29)$$

Detailed measurement of loading-rate dependency may be avoided by making a precise measurement of force distribution at a single loading rate, in case a wide range of loading-rate change is not accessible for the particular AFM instrument in use. The histogram of rupture force has a non-Gaussian shape which is approximated by the following equation [23] and used to deduce the value of $\Delta x$ from the rupture force histogram obtained at a single loading rate.

$$P\{f_{\text{rup}}\} = C\exp\{(f_{\text{rup}} - f^*)/f_\beta\}\exp \qquad (3.30)$$
$$[1 - \exp\{(f_{\text{rup}} - f^*)/f_\beta\}]$$

The loading-rate dependence is observed not only in unbinding reactions but also in all the cases where force is used to deform or disrupt the mechanical system.

## 3.9 FORCE CLAMP METHOD

It is also possible to use the force clamp method to obtain $\Delta x$ and $k_0$. In this method, an applied force is kept at a constant level for a prolonged duration of time, and the time duration from the beginning of force application to the time of bond rupture is observed [24, 25].

## 3.10 SPECIFIC VERSUS NONSPECIFIC FORCES

By using the force mode of AFM, one can measure the inter-action force between sample A on the probe and sample B on the substrate. The force may be an attractive or an repulsive force that starts working before the contact between A and B, or it may be the force required to separate A from B after the formation of a bimolecular complex between them. One difficulty in force measurement is that force is force whatever its origin, meaning that the probe senses all kinds of force working on it, whereas the experimentalists want to measure only one specific kind of force, *i.e.*, the interaction force between A and B. We call the force we aim to measure the 'specific force', and all other interaction forces are 'nonspecific' force or 'noise'. 'Specific' force is only based on the subjective choice of the experimentalist. The question is how to distinguish a specific force from nonspecific ones.

Ideally, one should establish the following observations to claim that the experimentalist is measuring the specific force.

- No interaction between the probe modified and treated exactly the same way as the probe to be modified with ligand B and the sample on the substrate.

- No interaction between the modified probe with B and the substrate itself.

- Observation of positive interaction force between a probe modified with B and sample A on the substrate.

- Addition of 'inhibitor' of the interaction between A and B to the sample solution to see whether the inhibitor specifically inhibits the interaction between A and B.

- Use of a force as small as possible in the initial contact of the probe with the sample, or use a crosslinker with a long spacer unit for the immobilization of B on the probe. This is to avoid pushing A on the substrate or B on the probe too hard so that they would not be damaged and would become the secondary source of adhesive interaction. In general, denatured proteins are stickier against almost any surfaces than native ones are, and thus pushing the sample proteins beyond their elastic limit increases the probability of unwanted nonspecific adhesion events.

Nonspecific interaction between the AFM probe and the substrate, if at all, is often revealed by the force curve as shown in Figure 3.9.

In the retraction regime of the force curve in Figure 3.9, the probe stays in contact with the substrate for a prolonged time and then suddenly gets detached from the substrate in a single step. This type of force curves likely represents nonspecific adhesion of the probe to the substrate.

The use of spacer molecules between the sample and the substrate and/or the probe is recommended. In such a case, the initial phase of the retraction regime of the force curve should be similar to that shown in Figure 3.10, where the force curve clearly has a nonlinear extension of the spacer before the probe is detached from the sample.

It is important to verify that the initial part has only a small or no 'triangular' adhesion part before the spacer extension starts. If the initial triangle adhesion is small and spacer extension part is clearly distinguished from it, the interaction force can be estimated from the rupture force after spacer extension, but it is important to exclude the initial one or two force peaks that are likely to involve nonspecific interactions.

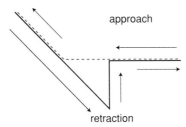

**Figure 3.9** Typical nonspecific adhesion-curves with a large triangular force-curve in the retraction regime.

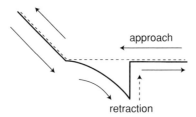

**Figure 3.10** Spacer is valuable when force curves of interaction are obtained. The rupture event that appears after an extension of a long spacer of known length most likely corresponds to the unbinding of the focused pair. The nonlinear increase of the loading force, however, complicates the calculation of the loading rate.

## Bibliography

[1] Ashkin, A. (1992), Forces of a single-beam gradient laser trap on a dielectric sphere in the ray optics regime, *Biophys. J.*, 61, 569–582.

[2] Kawata, S., Ohtsu, M., and Irie, M. (eds) (2002), 'Nano-optics', Chapter 4 Springer, Berlin, pp. 88–89.

[3] Ashkin, A., Schütze, K., Dziedzic, J. M., Euteneuer, U., and Schliwa, M. (1990), Force generation of organelle transport measured in vivo by an infrared laser trap, *Nature*, 348, 346–352.

[4] Hénon, S., Lenormand, G., Richert, A., and Gallet, F. (1999), A new determination of the shear modulus of the human erythrocyte membrane using optical tweezers, *Biophys. J.*, 76, 1145–1151.

[5] Binnig, G., Rohrer, H., Gerber, Ch., and Weibel, E. (1982), Tunneling through a controllable vacuum gap, *Appl. Phys. Lett.*, 40, 178–180.

[6] Binnig, G., Rohrer, H., Gerber, Ch., and Weibel, E. (1982), Surface Studies by Scanning Tunneling Microscopy, *Phys. Rev. Lett.*, 49, 57–61.

[7] Bonnell, D. (2000), 'Scanning Probe Microscopy and Spectroscopy: Theory, Techniques, and Applications', Wiley VCH New York, NY.

[8] Wiesendanger, R. (1994), 'Scanning Probe Microscopy and Spectroscopy', Cambridge University Press. Cambridge, UK.

[9] Binnig, G., Rohrer, H., Gerber, Ch., and Weibel, E. (1983), 7 × 7 Reconstruction on Si(111) resolved in real space, *Phys. Rev. Lett.*, 50, 120–123.

[10] Takayanagi, K., Tashiro, Y., Takahashi, M., and Takahashi, S. (1985), *J. Vac. Sci. Technol.*, A3 1502.

[11] Binnig, G., Quate, C. F., Gerber, Ch. (1986), Atomic force microscope, *Phys. Rev. Lett.*, 56, 930–933.

[12] Butt, H. J., Cappella, B., and Kappl, M. (2005), Force measurements with the atomic force microscope: technique, interpretation and applications, *Surf. Sci. Rep.*, 59, 1–152.

[13] Dill, K. A. and Bromberg, S. (2002), 'Molecular Driving Forces: Statistical Thermodynamics in Chemistry and Biology', Garland Science, New York, NY.

[14] Timoshenko, S. P. and Gere, J. M. (1972), 'Mechanics of Materials', (1st ed.), PWS Publishings & Co. Boston, MA.

[15] Giessble, F. J. (2002), Chapter 2 in 'Noncontact Atomic Force Microscopy' by Morita, S., Wiesendanger, R., and Meyer, E., Springer. Berlin Germany.

[16] Morita, S. (2002), Chapter 1 in 'Noncontact Atomic Force Microscopy' by Morita, S., Wiesendanger, R., and Meyer, E., Springer. Berlin Germany.

[17] Evans, E., Berk, D., and Leung, A. (1991), Detachment of agglutinin-bonded red blood cells. I. Forces to rupture molecular-point attachments, *Biophys. J.*, 59, 838–848.

[18] Evans, E., Ritchie, K., and Merkel, R. (1995), Sensitive force technique to probe molecular adhesion and structural linkages at biological interfaces, *Biophys. J.*, 68, 2580–2587.

[19] Walter, N., Selhuber, C., Kessler, H., and Spatz, J. P. (2006), Celluar unbinding forces of initial adhesion processes on nanopatterned surfaces probed with magnetic tweezers, *Nano Lett.*, 6, 398–402.

[20] Tan, J. L., Pirone, D. M., Gray, D. S., Bhadriraju, K., and Chen, C. S. (2003), Cells lying on a bed of microneedles: an approach to isolate mechanical force, *Proc. Natl. Acad. Sci. USA*, 100, 1484–1489.

[21] Fritz, J., Baller, M. K., Lang, H. P., Rothuizen, H., Vettiger, P., Meyer, E., Guntherodt, H.-J., Gerber, Ch., and Gimzewski, J. K. (2000), 'Translating Biomolecular Recognition into Nanomechanics', *Science*, 288, 316–318.

[22] Evans, E. and Ritchie, K. (1997), Dynamic strength of molecular adhesion bonds, *Biophys. J.*, 72, 1541–1555.

[23] Takeuchi, O., Miyakoshi, T., Taninaka, A., Tanaka, K., Cho, D., Fujita, M. et al. (2006), Dynamic-force spectroscopy measurement with precise force control using atomic-force microscopy probe, *J. Appl. Phys.*, 100, 074315–074320.

[24] Shao, J. Y. and Hochmuth, R. M. (1999), Mechanical anchoring strength of L-selectin, beta2 integrins, and CD45 to neutrophil cytoskeleton and membrane, *Biophys. J.*, 77, 587–596.

[25] Oberhauser, A. F., Hansma, P. K., Carrion-Vazquez, M., and Fernandez, J. M. (2001), Stepwise unfolding of titin under force-clamp atomic force microscopy, *Proc. Natl. Acad. Sci. USA.*, 98, 468–472.

## CHAPTER FOUR

# POLYMER CHAIN MECHANICS

## Contents

## 4.1 POLYMERS IN BIOLOGICAL WORLD

Biological structures are made of polymeric substances such as proteins, polysaccharides, and nucleic acids. Examples of such natural polymers are given below.

### I. Proteins

**Structural proteins:**  collagen, elastin, keratin, crystalline proteins, adhesive proteins

**Enzymes:** proteases, nucleases, glycosidases, lipases, esterases, dehydrogenases, oxygenases, carboxylases, synthetases

**Antibodies:** IgA, IgD, IgE, IgG, IgM

**Hormones:** growth hormones, insulin, glucagon

**Transport and storage proteins:** albumin, lipoproteins, ferritin, transferrin, hemoglobin, myoglobin

**Receptors and channels:** rhodopsin, insulin receptor, sodium channel, pottasium channel, anion channel

**Nucleo proteins:** histones

**Membrane proteins:** glycophorins, cadherins, integrins, stomatin, Band 3

**Matrix proteins:** laminin, fibronectin, vitronectin

## II. Polysaccharides

**Amylose and amylopectin:** polymers of glucose

**Cellulose:** polymers of glucose

**Chitin:** polymers of $N$-acetyl glucosamine

**Mannan:** polymers of mannose

**Galactan:** polymers of galactose

## III. Nucleic acids

### DNA
### RNA

Many proteins obtain sugar moieties after biosynthesis of polypeptide chains by the action of various kinds of enzymes, and the products are called glycoproteins. Addition of sugar chains of different lengths and different sequences, which are well defined for a given type of glycoprotein, enormously expands the surface variability of glycoproteins. Many membrane proteins are, in fact, glycoproteins conferring a wide range of variations of chemical and physical nature to the cell surface, which of course is biochemically exploited by the host cell. Mechanically, what is interesting is the numerical assessment of the rigidity, elasticity, yielding force, tensile strength, etc, of the native conformations of a variety of proteins as the fundamental material to support life from both structural and functional points of view.

## 4.2 POLYMER CHAINS

Proteins, nucleic acids, and polysaccharides are all polymeric substances and share fundamental properties with synthetic polymers. In this chapter, some basic ideas that are necessary for understanding nano–mechanical researches are introduced.

I. **Contour length:** It is the physical distance along the main chain of a polymer molecule from the $i$th to the $j$th segment $(L_{ij})$. The total contour length is the length from the first to the last segment $(L_0)$.

II. **End-to-end distance:** It is the bird's eye distance between the first and the last segment of a polymer chain. A statistical average of root mean squared end-to-end distance is $R = <h^2>^{1/2}$.

III. **Randomly coiled chain:** Segments are connected with universal joints having no preferred restrictions for its rotation around the solid angle of $4\pi$.

IV. **Persistence length:** It is the contour length from the $i$th to the $k$th segment where the directional correlation between two segmental vectors is lost.

V. **Radius of gyration:** It is the root-mean-squared distance weighted by the segmental mass of all the segments from the center of mass of a polymer chain.

VI. **Entropic elasticity:** It is the spring-like behavior of a polymer chain due to its entropic stability at equilibrium. It is manifested when its conformation is disturbed.

A polymer chain is basically a linear collection of $n$ small segments of equal length $L$. Each segment has two neighboring segments, except for the two end segments. A simple mathematical

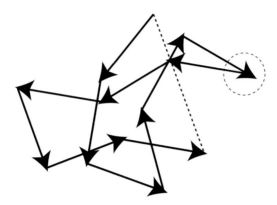

**Figure 4.1** A polymer chain model as a freely jointed chain (FJC). The rigid segment of length $L$ is linearly connected at their ends by a joint (circled) having no preference for its rotation over the entire solid angle of $4\pi$. The dotted line represents the end-to-end distance in this particular case.

model of such a polymer chain assumes that segments are connected with universal joints, with no restraints on the relative rotational freedom for the neighboring segments as illustrated in Figure 4.1.

This type of molecules has an interesting characteristics called rubber-like or entropic elasticity. Since, at each joint, the two independently variable angles between the neighboring segments, $\theta$ and $\phi$ (Figure 4.2), can assume any value between 0 and $\pi$ for the former and between 0 and $2\pi$ for the latter, the polymer has a large number of all possible conformations defined by different values of $\theta$ and $\phi$. Although there is no preferred values for the two angles, the degeneracy for different values of $\theta$ is different and proportional to $2\pi \sin\theta d\theta$, which is largest for $\theta = \pi/2$. Thus $< \theta >= \pi/2$. For larger or smaller values of $\theta$, the degeneracy is less, which means that the conformational entropy of the chain is at its maximum for states with $< \theta >= \pi/2$. The definition of the conformational entropy being $S = k_B \ln Z$, where $Z$ is the number of different states with the same value of $< \theta >$, and conformational states with either larger or smaller $< \theta >$ have both smaller $Z$ and $S$; thus these are less stable in terms of Gibbs energy

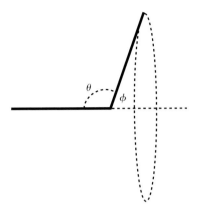

**Figure 4.2** Chemical bonds in a polyethylene chain have a fixed bond length $L$ and a bond angle $\theta$. A single bond has the freedom of rotation around its axis, which is characterized by the angle $\phi$.

($G = H + TS$, where $H$ and $T$ are, respectively, enthalpy and temperature in K).

When a polymer chain is stretched or compressed from the equilibrium state of maximum entropy by the application of an external force, it will exert a reaction force and when the applied force is released, it will snap back to the original state much as a spring. When the polymer deformation is small, the spring is Hookean, but as the deformation becomes large, the polymer spring will be highly nonlinear and non-Hookean. The spring-like nature of a randomly coiled polymer chain forms the basis of rubber elasticity as shown in Figure 4.3.

According to a statistical treatment of a polymer chain, the distribution function of its segments has a Gaussian form as given below.

$$P(R) = \left( \frac{3}{2\pi nb^2} \right)^{3/2} \exp\left( -\frac{3R^2}{2nb^2} \right) \tag{4.1}$$

Thus, the spring constant of a polymer chain as it is displaced from the equilibrium conformation is $3k_B T/nb^2$, because force $F = k_B T \partial \ln P(R)/\partial R = (3k_B T/nb^2)R$.

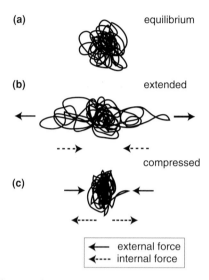

**Figure 4.3** A polymer chain has a property of a spring. When stretched (b) or compressed (c) from its equilibrium conformation in (a), it exerts a reaction force to return to the original equilibrium state. Rubber is a typical elastic material and its elasticity is based on the spring-like nature of polymer chains. A simple collection of polymer chains may not behave as an effective spring but, when they are cross-linked similarly to vulcanized rubber, an almost perfectly elastic spring emerges.

## 4.3 END-TO-END DISTANCE

### 4.3.1 Definition

Polymers are made of a large number of covalently bonded monomeric units that are usually aligned in a linear chain but sometimes with branching out into multiple chains. The monomeric units may be similar from one end of the molecule to the other or may be different with or without regularity. One of the simplest example is polyethylene, which is entirely made of linear sequence of methylene groups ($-CH_2-CH_2-$), except for the two end groups which are methyl groups. It is not soluble in water but can be dissolved in aromatic hydrocarbons (*i.e.*, toluene and xylene) or chlorinated solvents (*i.e.*, trichloroethane and trichlorobenzene) at elevated temperatures. In a solubilized state, polyethylene does not have a fixed structure, but has a somewhat globular conformation with some solvent molecules inside the globule. The covalent bonds between each $CH_2$ can rotate with three most stable configurations, namely, one *trans* and two gauche forms $(\pm)$, at a lower temperature but can almost freely rotate at a moderately high temperature. If each single bond finds three stable positions, there are $3^2 = 9$ allowed configurations for three monomeric units separated by one single bond, $3^3 = 27$ for four units, $3^{100} \sim 5.14 \times 10^{47}$ for 101 units, etc. Quite a large number of different conformations can be assigned to a single chain of polyethylene of $n$ single bonds, if $n$ is large. One of them is a unique conformation where all the bonds are in the *trans* configuration; thus the chain assumes the most extended conformation for a given polyethylene molecule with the longest end-to-end distance, *i.e.*, the distance between the one end of the molecule to the other end.

The longest end-to-end distance coincides with the contour length, *i.e.*, the length of the chain as measured along it. All other conformations having a mixture of trans and $\pm$-gauche configurations have shorter end-to-end distances, because the presence of $\pm$-gauche configurations with respect to a particular single bond brings the two methylene units separated by three bonds to come closer to each other so that the chain makes an inward turn compared with when the particular bond is in *trans* configuration. If $n_t$ and $n_g$ bonds out of the total of $n$-bonds are, respectively, in *trans* and $\pm$-gauche configurations, there are $N = n!/(n_t!n_g!)$ different ways to distribute *trans* and gauche bonds in one chain. This is quite a large number when both $n_t$ and $n_g$ are large, and consequently, in the population of the molecules, those having equally large number of trans and gauche bond configurations predominate and the population average of the end-to-end distance is much shorter than the contour length of the molecule.

## 4.3.2 Randomly coiled polymer

A polymer chain having all the main chain chemical bonds with unrestricted freedom of rotation is called a random coil or randomly coiled molecule. As explained in the previous section, such molecules have much shorter end-to-end distance than their contour length meaning they are coiled up into a globule with a diameter in the order of the end-to-end distance. Since the molecules are dissolved in solvents, affinity between monomeric units of the polymer and the solvent molecules is reasonably high, allowing solvent molecules to almost freely solvate polymer monomeric units inside and outside the globular conformation. Within the domain of polymer globule, there are a significant number of solvent molecules, and they tend to move with polymer molecules as the latter move in the solvent due to diffusion or under other applied force such as electrophoretic force. There are several different models to represent characteristic behaviors of randomly coiled polymers.

### 4.3.3 The FJC (Freely Jointed Chain)

Polymer chains can be approximated by several different models, each emphasizing a particular aspect of the polymer conformation. An often quoted one is the freely jointed chain where a polymer molecule is modeled as composed of $N$ rigid straight rods which may encompass several physical segments and are joined at the two ends with completely freely rotatable universal joints. In this case, the end-to-end distance vector $R$ is the sum of all the newly defined segmental vectors, $r_i$, and thus, $R = \sum_{i=1}^{i=N} r_i$.

Because the polymer chain is constantly changing its conformation, $i.e.$, the angle between the two segments is constantly changing, taking all the values of $\theta$ and $\phi$ with equal probability, the average end-to-end distance should be calculated by summing segmental vectors $r_i$ over all possible intersegment angles (Figure 4.2)

Thus,

$$< R > = \left\langle \sum_{i=1}^{i=N} r_i \right\rangle = \sum_{i=1}^{i=N} < r_i > = 0 \qquad (4.2)$$

where the quantities in $<\ >$ represent time-averaged values over all the possible conformations for a single polymer molecule, which is equivalent to the average over a large number of molecules at a fixed time. Since there is no correlation between the direction of two segmental vectors, the average of the sum of all the segmental vectors approaches zero as the number of segments becomes large and/or the time for averaging becomes longer. Then, we take the average of the square of the end-to-end vector as follows.

$$\boldsymbol{R}^2 = \left(\sum_{i=1}^{i=N} \boldsymbol{r}_i\right)^2 = \sum_i \boldsymbol{r}_i^2 + 2\sum\sum_{i \neq j} \boldsymbol{r}_i \cdot \boldsymbol{r}_{i+j} \qquad (4.3)$$

The first term on the right-hand side of Eq. (4) is the sum of the square of each of the segmental vectors, which is equal to $N$ times the square of the segmental length. The second term becomes vanishingly small compared with the first term as $N$ becomes large. Thus,

$$< \boldsymbol{R}^2 > = \sum < \boldsymbol{r}_i^2 > + 2\sum\sum < \boldsymbol{r}_i \cdot \boldsymbol{r}_{i+j} > = N < \boldsymbol{r}_i^2 > \qquad (4.4)$$

$$< \boldsymbol{R}^2 > = Nb^2 \quad \text{where } b = |r_i| \qquad (4.5)$$

As implicated above, it is possible to fit a real polymer chain to the freely jointed chain model by not considering the segmental vector as equivalent to the monomeric unit, but assuming that the chain is made of freely jointed segments of length $L_K$, which may be close to the length of the monomeric unit or much longer depending on the stiffness of the polymer chain.

$$L_K \equiv \lim_{L \to \infty} \frac{< R^2 >}{L} \qquad (4.6)$$

The $L_K$ defined as above is called Kuhn's statistical segment, emphasizing the fact that it is not equivalent to the actual segment that makes up the polymer chain. A large number of different conformations having different intersegmental angles would give a similar value of the end-to-end distance, but there are fewer of them giving extremely large or small values. Thus, the number of different conformations that give the same value of end-to-end distance has a Gaussian distribution as given below.

$$\Phi(R, N) = \left(\frac{3}{2\pi N L_K^2}\right)^{3/2} \exp\left(\frac{-3 R^2}{2 N L_K^2}\right) \qquad (4.7)$$

The chain is thus often referred to as the Gaussian chain. Because there are so many different conformations that give not so long or not so short end-to-end distance, most often the chain has an intermediate end-to-end distance, and if it is either stretched or compressed beyond the most favored conformation by the application of an external force, it tries to return to the original state like a spring. To calculate the reaction force, we start by calculating the value of the conformational entropy, $S$, which is proportional to the logarithm of the number of different conformational states that give the same observable, in this case, end-to-end distance.

$$S(R, N) \propto k_B \ln(\Phi(R, N)), \quad \text{therefore,} \quad S = S_0 - \frac{3k_B R^2}{2 N L_K^2}$$
$$(4.8)$$

Next, by using the value of entropy we obtained above, the Helmholtz free energy $\Phi$ is obtained as below.

$$\Phi = E - TS = F_0 + \frac{3k_B T R^2}{2 N L_K^2} \qquad (4.9)$$

The force is defined as the first derivative of the Helmholtz free energy with − sign.

$$F = -\frac{d\Phi}{dR} = \frac{3k_B T R}{N L_K^2} \qquad (4.10)$$

We learned from the above result that the force to recover the original end-to-end distance is proportional to the degree of extension or compression $R$; thus a polymer chain behaves like a Hookean spring with a spring constant of $3k_B T / N L_k^2$.

This behavior can be observed in the stretching and compressing experiment done on a single chain of randomly coiled chain.

## 4.4 PERSISTENCE LENGTH

The persistence length of a polymer chain is a measure of its stiffness. It is defined as the length over which correlations in the direction of the tangent of the polymer chain at one point are lost. If the angle between the $m$th and $l$th segments is $\theta_{ml}$ for two segments within a short distance, it is closer to $\pi$ compared with those at a large distance. $\theta_{ml}$ decays from $\pi$ (perfect correlation) to $\pi/2$ (no correlation). Thus, the directional correlation of two segments exponentially depends on the contour distance $L_{ml}$ with a characteristic decay constant of $p$, which is the definition of the persistence length.

$$< \cos(\theta_{ml}) > = \exp\left( -\frac{L_{ml}}{p} \right) \qquad (4.11)$$

It is possible to prove that the persistence length is one half of the Kuhn statistical length of a freely joined chain.

A polymer chain with a short $p$ is more flexible than the one with a longer $p$, meaning the former has a much larger conformational state as a random coil than the latter, provided that they have the same contour length $L = np$, where $n$ is the number of the segments having the length equal to the persistence length. Consequently, the coiled state of the former polymer has larger entropy and requires a larger tensile force to be extended than the latter.

The relationship between the tensile force $F$ and the chain extension $L$ can be obtained from the consideration of the potential energy of each segment under a uniaxial tensile force. Let us consider a segment at an angle $\theta$ in a tensile force field of $F$ to $x$ and $-x$ direction. The potential energy of the segment is equal to $-Fb\cos\theta$ and the corresponding Boltzmann factor is $\exp[Fb\cos\theta/k_B T]$. The average length of the segment in the direction of the force field is

$$b < \cos\theta > = \frac{\int\limits_{\theta=0}^{\theta=\pi} b\cos\theta e^{Fb\cos\theta/k_B T}d\tau}{\int\limits_{\theta=0}^{\theta=\pi} e^{Fb\cos\theta/k_B T}d\tau} \tag{4.12}$$

$$\text{where } d\tau = 2\pi b \sin\theta d\theta \tag{4.13}$$

By setting $Fb\cos\theta/k_B T = y$, we find first, $d\tau = -2\pi (k_B T/F)dy$.
The denominator of Eq. (4.12) is

$$(-2\pi)\left(\frac{k_B T}{F}\right)\int_A^{-A} e^y dy = (-2\pi)\frac{k_B T}{F}(e^{-A} - e^A)$$

$$\text{where } \quad A = \frac{Fb}{k_B T} \tag{4.14}$$

The multiplier is

$$(-2\pi)\left(\frac{k_B T}{F}\right)^2 \int_A^{-A} y e^y dy \tag{4.15}$$

$$= (-2\pi)\left(\frac{k_B T}{F}\right)^2 \left[-A(e^{-A} + e^A) - (e^{-A} - e^A)\right] \tag{4.16}$$

Thus, the integral is

$$\left(\frac{k_B T}{F}\right)\left[\frac{-A(e^{-A} + e^A) - (e^{-A} - e^A)}{e^{-A} - e^A}\right] \tag{4.17}$$

$$= b\left(\frac{e^A + e^{-A}}{e^A - e^{-A}}\right) - \frac{k_B T}{F} \tag{4.18}$$

$$= b\left(\coth A - \frac{1}{A}\right) \tag{4.19}$$

By dividing both sides of the above equation by $b$, we obtain the expression for $< b\cos\theta/b >$, which is equal to the relative extension of the chain against its total contour length, $L_0 = Nb$, when multiplied by the total number of segments, $N$. Thus,

$$\frac{L}{L_0} = \coth \frac{Fb}{k_B T} - \frac{1}{\frac{Fb}{k_B T}} \tag{4.20}$$

$$= \mathcal{L}\left(\frac{Fb}{k_B T}\right) \tag{4.21}$$

where $\mathcal{L}(x) = \coth x - 1/x$ is the Langevin function. By using the inverse Langevin function, the force is given in terms of relative extension.

$$\frac{Fb}{k_B T} = \mathcal{L}^{-1}\left(\frac{L}{L_0}\right) \tag{4.22}$$

The force versus extension curve has been simulated as the following equation [1], where the segment length $b$ is replaced by the persistence length $p$ by keeping the total contour length $L_0$ unchanged.

$$F = \frac{k_B T}{p}\left[\frac{1}{4}\left(1 - \frac{L}{L_0}\right)^{-2} - \frac{1}{4} + \frac{L}{L_0}\right] \tag{4.23}$$

When changes in covalent bonds must be considered at a longer extension, the following fitting curve has been postulated [1]. In general, when the tensile force reaches 1 nN, an opening of bond angles beyond the equilibrium must be considered and when the force approaches 2 nN, extension of bond length comes to play an important role.

$$F = \frac{k_B T}{p}\left[\frac{1}{4}(1 - Z)^{-2} - \frac{1}{4} + Z\right], \quad \text{where } Z = \frac{L}{L_0} - \frac{F}{K_0} \tag{4.24}$$

### 4.4.1 Effect of cross-links

Many protein molecules have intramolecular cross-links (disulfide bonds), and the effect of such crosslinks was discussed in relation to the mechanical properties of the chain [2]. Cross-links inside a protein and other polymers are expected to make them more rigid with the rigidity modulus $G = \rho RT/M$ where $\rho$ and $M$ are, respectively, the density of the material and the average molecular weight of the material between successive cross-links. The effect should be tested by using compression experiment on an AFM.

## 4.5 POLYMERS IN SOLUTION

### 4.5.1 General cases

When polymer molecules are dissolved in good solvent where the affinity between polymer segments and the solvent molecule is higher than solvent–solvent or segment–segment affinities, relatively expanded conformations with larger end-to-end distance are more common. With the decrease of the segment–solvent affinity, the polymer molecules start to shrink with the decrease of the end-to-end distance. Since molecules with large end-to-end distance occupy a larger volume, they give rise to a solution of high viscosity. Thus, the state of expansion of a polymer molecule can be estimated from the measurement of the intrinsic viscosity, $[\eta]$, according to the relation given by Flory [3]. In the following equation, $\Phi$ and $< h^2 >$ are, respectively, a characteristic constant (or Flory constant $= 2.8 \times 10^{23}\,\text{mol}^{-1}$) and the statistical average of square of the end-to-end distance. Since the intrinsic viscosity has the dimension of $\text{m}^3/\text{kg}$ (effective volume per unit mass), it is a measure of molecular expansion of a polymer chain.

$$[\eta] = \Phi \frac{< h^2 >^{3/2}}{M} \qquad (4.25)$$

For denatured proteins, the end-to-end distance was estimated based on the relationship reported by Tanford [4]

$$M_0[\eta] = 77.3n^{0.666} \tag{4.26}$$

where $M_0$ and $n$ are the molecular weight of the average residue and the number of amino acid residues, respectively.

Determination of the end-to-end distance of polymer molecules becomes important to obtain an estimate of the persistence length, which is defined as the contour length of the polymer chain over which the tangential direction becomes uncorrelated.

The radius of gyration, $R_g$, is also a good measure of polymer expansion in solution, which can be calculated from the result of laser light scattering [5].

### 4.5.2 Denatured proteins and DNA

The native conformation of proteins and nucleic acids can be disrupted without breaking their covalent structures and, in many cases, the resulting molecules have conformations very similar to randomly coiled polymers. Once the denaturing conditions are eliminated, the original native conformation can be restored. In the denaturation process, most or all of the intersegmental non-covalent interactions are reduced to the level of thermal energy so that the polypeptide chain behaves like a thermally fluctuating randomly coiled chain. All the amino acid residues expose their side chains to the solvent, regardless of their hydrophobicity or hydrogen bonding propensity. It is rather amazing that the native conformation can be restored within a short time after the reversal of the environmental parameters because it is the choice of one out of $10^{20}$ or more of possible other conformations.

The solution studies of protein denaturation provided detailed knowledge of the thermodynamics and cooperativity of inter-segmental interactions, but little about the rigidity of protein molecules as a whole and its local variability within a molecule.

## 4.6 POLYMERS ON THE SURFACE

Investigation of polymer molecules adsorbed on a solid surface has been a focus of attention and some interesting work has been done at the single molecular level using AFM. Synthetic polymers are usually randomly coiled and rather uniform in the chemical nature along its contour length, and adhesion to the surface at some specific part(s) of the chain is usually not observed, except in the case of block copolymers. Instead, pulling mechanics of a polymer chain from the solid surface is often characterized by the appearance of a force plateau, which is interpreted as the continuous de-adhesion process of a stretched chain on the substrate.

When polymer molecules are chemically end-grafted on the surface, they adopt a shape called 'mushroom' when the number density on the surface is small and, therefore, there is an ample space for each polymer molecule to adopt an extended conformation. The height of the mushroom is close to one-half of the end-to-end distance of the free polymer under the same conditions, because it is immobilized at one of its ends. As the number density on the substrate increases, the space allottable to each polymer molecule becomes smaller and the molecules become laterally compressed and eventually form vertically elongated random coils because the polymer chains do not easily mix their segments. Such alignment of elongated polymer is often called 'brush'.

Haupt et al. reported the imaging and force measurement of the mushroom and the brush states of the polymer graft on a solid surface [6]. Imaging of the mushroom state is an interesting attempt since the soft mushroom tends to escape from the AFM probe as the latter approaches them. The behavior of the mushroom is also interesting as a model of denatured protein on a solid surface as studied by Afrin et al. [7]. The Young's modulus of randomly coiled polymer has been reported to be in the range of 1–5 MPa, a similar value to rubber or denatured protein.

 ## 4.7 POLYMERS AS BIOMIMETIC MATERIALS

Replacement of tissues and organs in patients is becoming a reality in medical and medical engineering field. To construct artificial tissues and organs, effort has been made for the development of polymeric, ceramic, and metallic materials with good biocompatibility. The major concern from the medical side is the biocompatibility including biodegradatability. Proteins have a tendency to adsorb to synthetic surfaces, and they irreversibly adhere and aggregate and often accelerate blood coagulation, which is a serious threat to the health of the patients.

 ## 4.8 Polymer Pull-out

Pulling out a polymer chain from a solid polymer surface has been studied both experimentally and theoretically because it plays an important role in the lubrication and friction problems involving a solid polymer interface. In one such experiment shown in Figure 4.4,

**(a)** extended contacts

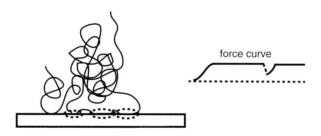

**(b)** point contact

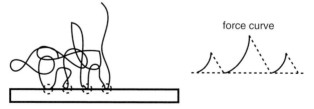

**Figure 4.4**    A polymer chain is pulled out of the monolayer on a solid surface. In (a), a polymer chain is adhered to the surface for a certain extension, and the force curve has a plateau at a constant force corresponding to the desorption force. In case (b), a polymer chain is adhered to the solid surface at several discreet points, with flexible chain segments connecting them. The force curve in this case is a series of sharp force peaks called a sawtooth pattern.

a polymer chain (trimethyl terminated polydimethylsiloxane (PDMS, chemical formula $(CH_3)_3\text{-Si-O}(\text{-Si-}(CH_3)_2\text{-O})_n\text{-Si-}(CH_3)_3)$) was pulled out of the polymer monolayer on a silicon surface in air as well as in water, both providing poor solvent conditions for the polymer.

The resulting force curves showed long plateau forces, most likely representing uncoiling globules in poor solvents. In poor solvents, a polymer chain is expected to be in globular coil, and extension from the coiled state to an elongated state is predicted to be accompanied by a plateau force. Such a behavior has been predicted by theory [8] and confirmed by an experiment using a polyelectrolyte chain [6]. There is a transition between the collapsed coiled state and the extended brush state, and this transition was recorded by changing the imaging force of polyelectrolyte under different solvent conditions [9, 10]. This behavior in forced extension in poor solvents is very different from the extension mechanics of randomly coiled polymer in good solvents. The appearance of a plateau force in the pulling experiment of a collapsed chain means that the same level of intersegmental interaction force of approximately 25 pN in air and 50 pN (at a pulling speed of 1 $\mu$m/s) in water is successively ruptured during the pulling process. Explanation based on the difference in surface energy between the polymer and the solvent is given.

## Bibliography

[1] Smith, S. B., Cui, Y., and Bustamante, C. (1996), Overstretching B-DNA: the elastic response of individual double-stranded and single-stranded DNA molecules, *Science*, 271, 795–799.

[2] Makarov, D. E. and Rodin, G. J. (2002), Configurational entropy and mechanical properties of cross-linked polymer chains: implications for protein and RNA folding, *Physical Review E*, 66, 011908.

[3] Flory, P. J. (1953), 'Principles of Polymer Chemistry', Cornell University Press. Ithaca, NY.

[4] Tanford, C., Kawahara, K., and Lapanje, S. (1967), *J. Am. Chem. Soc.*, 89, 729–735.

[5] Cantor, C. R. and Schimmel, P. R. (1980), 'Biophysical Chemistry Techniques for the Study of Biological Structure and Function', W. H. Freeman, San Francisco.

[6] Haupt, B. J., Senden, T. J., and Sevick, E. M. (2002), AFM evidence of Rayleigh instability in single polymer chains, *Langmuir*, 18, 2174–2182.

[7] Afrin, R., Alam, M. T., and Ikai, A. (2005), Pretransition and progressive softening of bovine carbonic anhydrase II as probed by single molecule atomic force microscopy, *Protein Sci.*, 14, 1447–1457.

[8] Halperin, A. E. and Zhulina, B. (1991), On the deformation behaviour of collapsed polymers, *Europhys. Lett.*, 15, 417–421.

[9] Farhan, T., Azzaroni, O., and Huck, W. T. S. (2005), AFM study of cationically charged polymer brushes: switching between soft and hard matter, *Soft Matter*, 1, 6668.

[10] Koutsos, V., Haschke, H., Miles, M. J., and Madani, F. (2003), Pulling single chains out of a collapsed polymer monolayer in bad-solvent conditions, *Mat. Res. Soc. Symp. Proc.*, 734, B1.6.1–B1.6.5.

## CHAPTER FIVE

# INTERACTION FORCES

---

## Contents

---

## 5.1 COVALENT VERSUS NONCOVALENT FORCE

Atoms in a molecule are bonded to each other through covalent bonds, which are generally quite strong and are responsible for the existence of most of the objects in the material world. Two atoms with vacancy in the valence orbitals, when approaching each other, either repel or become bonded depending on the nature of the spin state of the valence electrons. On approaching each other, their outermost orbitals interact to form two new orbitals whose energy levels are split into two, one lower and the other higher than the original ones. If the electrons in the original orbitals have opposite spins, they share a new orbital with lower energy, thus forming a stable covalent bond between the two atoms. The dependence of the potential energy on the distance between two atoms is often expressed in terms of Morse function as below.

$$V(r) = D[1 - \exp\{-a(r - r_{eq})\}]^2 \qquad (5.1)$$

where $D$ and $r_{eq}$ are, respectively, the dissociation energy and the equilibrium distance between two atoms. $D$ is in the range of a few hundred kJ/mol, and the force required to break them is in the order of a few nano-newtons under commonly employed experimental conditions of AFM. Under ordinary biochemical experimental conditions, covalent bonds are assumed unbreakable unless catalysts are used, but in mechanical experiments at the single molecular level, they are rather easily broken with a force about ten times larger than that required to break noncovalent interactions. It is, therefore, possible that a cluster of ten or more noncovalent interactions would be as strong as or stronger than a single covalent bond under mechanically stressed conditions.

## 5.2 BASICS OF ELECTROSTATIC INTERACTION FORCE

Atoms and molecules exert repulsive or attractive forces against each other. A simple example is a pair of permanently charged ions. If the charges on the pair are opposite in sign, they attract each other, whereas if they have charges with the same sign, they repel each other. This is called Coulombic interaction. Both attractive and repulsive forces obey the same law describing the dependence on the distance between the two ions. In the following equation, $V, q_1, q_2, \varepsilon_0, \varepsilon_r$, and $r$ are the potential, electric charges on molecules 1 and 2, respectively, the electric permittivity of vacuum ($8.85 \times 10^{-12}$ farad/m), the relative dielectric constant of the medium, and the distance between the two ions [1]. If the sign of $F$, the force between two charges, is negative, it represents an attractive force and vice versa.

$$V = \frac{q_1 q_2}{4\pi\varepsilon_0\varepsilon_r r}, \quad F = \frac{q_1 q_2}{4\pi\varepsilon_0\varepsilon_r r^2} \tag{5.2}$$

In an aqueous solution, the Coulombic interaction is weakened, first, by a large dielectric constant of water ($\varepsilon_r \sim 80$) and, second, by the presence of counter ions as described by the Debye–Hückel screening effect. An approximation of the screening effect in dilute salt solutions reduces the effective distance of the interaction by an exponential factor, $\exp[-\kappa r]$, as given below.

$$V = \frac{ze}{4\pi\varepsilon_0\varepsilon_r} \left[ \frac{\exp(\kappa a)}{1 + \kappa a} \frac{\exp(-\kappa r)}{r} \right] \tag{5.3}$$

where, $z, e,$ and $a$ are, respectively, the charge number, the elementary charge ($1.6 \times 10^{-19}$C), and the radius of the ion. In addition, $\kappa$ is defined as below and is called Debye–Hückel screening factor.

$$\kappa^2 = \frac{e^2 \sum n_i^0 z_i^2}{\varepsilon_0\varepsilon_r \kappa_B T} \tag{5.4}$$

where $z_i$, and $n_i^0$ are, respectively, the charge number of $i$th ion, and the bulk concentration of the $i$th ion. This factor is called screening factor because at a distance, where $r = 1/\kappa$, the original Coulombic interaction potential drops to $1/e$. $\kappa$ is used in many theories of interaction in an ionic solution.

If molecules are not charged, they are electrically neutral. Atoms and molecules are made of positively charged protons, uncharged neutrons, and negatively charged electrons. Neutral molecules have exactly the same number of protons and electrons. Atoms are basically spherical and electrically neutral, with the center of positive charge concurring with that of the negative charge, at least when averaged for a duration of time.

In some neutral molecules, the centers of positive and negative charges are coincident, but in some others they do not coincide. The former type of molecules are called nonpolar and the latter type polar. Examples of nonpolar molecules are $H_2$, $N_2$, $O_2$, $CH_4$, $C_2H_6$, $C_3H_8$, benzene, and other hydrocarbons, whereas $H_2O$, $CO_2$, $CO$, $CH_3COOH$, and $CH_3CH_2OH$ are examples of polar molecules. Polar molecules have atoms of significantly different 'electronegativity', meaning the tendency of atoms in a molecule to attract valence electrons close to their nuclei. Valence electrons reside for a longer time around the nucleus of atoms with a higher electronegativity than in the neighborhood of those having a lower electronegativity.

In polar molecules, we can define a dipole moment by placing one positive $(+q_1)$ and one negative $(-q_1)$ charge at a short distance of $d$, and the magnitude of a dipole, $\mu$, is defined as $\mu = qd$. A polar molecule has a permanent dipole of various strengths.

## 5.3 VARIOUS TYPES OF NONCOVALENT FORCES

Thus, the noncovalent interaction is classified into the following categories for convenience. All of them, in principle, are electrostatic interactions except for the last one, *i.e.*, hydrophobic interaction.

### 5.3.1 Charge–charge interaction

The electrostatic interaction is still effective in solutions, although its magnitude is considerably reduced due to the shielding effect explained by Debye and Hückel. Similar to the case of a free ion in a salt solution, to an electrically charged or polarized surface, counter ions of opposite charges accumulate, whereas those with the same charge (co-ions) are depleted. Many surfaces immersed in an aqueous salt solution are charged or polarized and show attractive or repulsive interactions with other surfaces, ions, and molecules in solution. The subject is treated by Derjaguin et al.

and is known as the DLVO theory [2].

$$V_{c-c} = Z^2 \lambda_{\mathrm{B}} \left( \frac{\exp(\kappa a)}{1 + \kappa a} \right)^2 \frac{\exp(-\kappa r)}{r}, \text{ where } \lambda_{\mathrm{B}} = \frac{e^2}{4\pi \varepsilon_0 \varepsilon_r k_{\mathrm{B}} T}$$
$$(5.5)$$

where $\lambda_{\mathrm{B}}$ is called the Bjerrum length, and $\kappa$ is the Debye–Hückel screening factor, the reciprocal of which, i.e., $\kappa^{-1}$, is called the screening length. The Bjerrum length is the separation at which the electrostatic interaction between two elementary charges is comparable in magnitude to the thermal energy scale, $k_{\mathrm{B}} T$.

### 5.3.2 Charge–dipole interaction

Looking at the situation where a fully charged positive ion approaches a polar molecule, we find that the latter presents its surface closer to the negative-charge center to the former, resulting in an overall attractive interaction.

$$V_{c-d} = -\frac{(ze)\mu \cos\theta}{4\pi \varepsilon_0 \varepsilon_r}, \text{ where } \mu = |\boldsymbol{\mu}| \qquad (5.6)$$

This interaction is strong enough to keep the concerned members in a fixed angle, $\theta$, and accounts for the solvation of an ion in polar solvents.

### 5.3.3 Dipole–dipole interaction

As indicated previously, when two polar molecules approach each other, they present their oppositely charged surfaces more closer and more often than similarly charged ones, again resulting in an overall attractive interaction. Dipole–dipole interaction is calculated by placing two dipoles of magnitudes, $\mu_1$ and $\mu_2$, at a distance of $r$, which is significantly larger than $d$, $(d \ll r)$. The two dipolar molecules in a given configuration are illustrated in Figure 5.1.

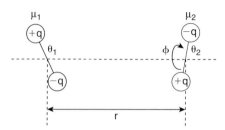

**Figure 5.1** Two dipoles are placed with a center-to-center distance of $r$ in an arbitrary orientation.

The result is as follows, with $\theta_1, \theta_2,$ and $\phi$ defined as in Figure 5.1.

$$< V_{d-d} >= -\frac{\mu_1 \mu_2}{4\pi\varepsilon_0\varepsilon_r r^3}[2\cos\theta_1 \cos\theta_2 - \sin\theta_1 \sin\theta_2 \cos\phi]$$

$$(5.7)$$

In a solution at room temperature, the interaction is weak compared with the thermal energy so that the two dipolar molecules tumble around each other, changing the angles $\theta_1, \theta_2,$ and $\phi$. It is necessary, therefore, to take an average energy for all possible orientations with the Boltzmann weighting factor keeping the distance $r$ fixed, and the result is

$$< V(r) >= -\frac{\mu_1^2 \mu_2^2}{3(4\pi\varepsilon_0\varepsilon_r)^2 k_B T r^6} \text{ for } k_B T > \frac{\mu_1 \mu_2}{4\pi\varepsilon_0\varepsilon_r r^3} \quad (5.8)$$

## 5.3.4 Dipole–induced dipole interaction

When a dipolar molecule and a nonpolar molecule approach each other, the charge distribution inside the nonpolar molecule experiences a rapid change so that the overall interaction between the two molecules becomes attractive. The interaction energy is dependent on the magnitude of the dipole of the polar molecule, $\mu$, and, importantly, on how easily the charge can redistribute inside the nonpolar molecule, which is quantified as the polarizability $\alpha_0$.

Again after averaging over all the possible orientations at a finite temperature, $T$, we obtain the interaction potential as

$$< V_{d-n} > = \frac{-\mu^2 \alpha_0}{(4\pi\varepsilon_0\varepsilon_r)^2 r^6} \tag{5.9}$$

## 5.3.5 Dispersion interaction

What happens if two nonpolar atoms or molecules approach each other? Although there was very little reason to expect either attractive or repulsive interactions between the two neutral atoms or molecules, experimentally a definite attractive interaction was noted. This interaction was explained by London based on the quantum mechanical theory. According to the theory, the distance, $d$, between the centers of positive and negative charges in a non-polar molecule or an atom, are very rapidly changing around the value of zero due to the orbital motion of electrons around the nucleus. The heavy nucleus cannot follow the rapid motion of electrons. Of course, the time average distance is zero in accordance with the assumption that the molecule is nonpolar within the measurement time. The fluctuation of $d$ of one molecule is transmitted to the other in the form of a photon (electromagnetic wave) and induces redistribution of charge density in the latter. The resulting effect is again an attractive interaction energy whose magnitude depends on the polarizability of the two nonpolar molecules, $\alpha_{01}$ and $\alpha_{02}$.

$$V_{n-n} = -\frac{3}{2}\frac{\alpha_{01}\alpha_{02}}{(4\pi\varepsilon_0\varepsilon_r)^2 r^6}\frac{h\nu_1\nu_2}{(\nu_1 + \nu_2)} = -\frac{3}{2}\frac{\alpha_{01}\alpha_{02}}{(4\pi\varepsilon_0\varepsilon_r)^2 r^6}\frac{I_1 I_2}{I_1 + I_2} \tag{5.10}$$

where $h$ is Planck's constant, and $\nu_1, \nu_2$, and $I_1, I_2$ are, respectively, the orbiting frequency of the electrons and the first ionization potential of molecule 1 and 2.

Thus, all interactions having the inverse 6th power over the distance are lumped as a single attractive interaction and termed 'van der Waals' interaction. The force being the first derivative of the potential, has the inverse 7th power dependence over the distance. The above derivation is based on the assumption that the size of the interaction particles are much smaller than the distance between them, *i.e.*, basically interactions between atoms and small molecules. Under actual experimental conditions of nanomechanics by using AFM, where the probe and the sample have finite sizes, all the interaction potentials must be summed over all pair-wise interactions as described in Ref. [1]. It should also be pointed out that the van der Waals interaction is affected in the presence of solvents because dipole–dipole interaction is basically electrostatic and is affected by the dielectric constant of water.

An example requiring such integration is given in Figure 5.2. In many cases where applicable, the probe sample interaction force in AFM experiments may be assumed to have an inverse 2nd power on the distance between them [3].

## 5.3.6 Hydrogen-bonding interaction

This is an attractive interaction between electronegative atoms in molecules when a hydrogen atom works as a bridge between them [4]. Examples can be found between N and O atoms in $-NH_2$ and $-CO$ groups, two O atoms in $-OH$ and $-CO$ groups, or two N atoms in $\equiv N$ and $-NH_2$ groups, etc., of different molecules or within the same molecule. It is not easy to express the distance dependence of the hydrogen bond.

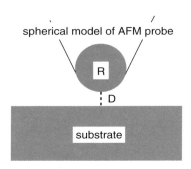

spherical model of AFM probe

**Figure 5.2** A macroscopic representation of sample–probe interaction requires integration of Lennard–Jones potential between all the pairwise interactions between the probe and the sample surface.

## 5.3.7 Hydrophobic interaction

It was noted that alkyl or aromatic groups in aqueous solution tend to segregate themselves out of aqueous solutions to form clusters among themselves [5, 6]. This effect is termed hydrophobic interactions; however, the physical explanation for the phenomenon has been and still is controversial to some extent. It is generally agreed that it is entropically favorable for two hydrophobic groups to come together resulting in a reduction of the total interface area with water. This is based on the experimental and theoretical assessment that water molecules around alkanes and aromatics are more ordered than in bulk water, thereby having a lower entropy compared with ordinary water. Reducing such interfaces by segregating out alkanes and aromatics does increase the overall entropy, and thus lowers the Gibbs energy of the solution as a whole. The two molecules that come into association due to hydrophobic interactions must be very close to each other, say within less than one nm but a formula for the distance dependence of the effect is still not obtained.

A controversy exists on the so-called long range hydrophobic effect. Some researchers have detected an attractive force between two hydrophobic surfaces placed in water at a distance of ten nanometers or even longer [7, 8]. At the present moment, the long-range hydrophobic effects are observed between two surfaces modified with amphiphilic surfactant molecules, but the relevance to biological phenomena is not clear.

The intramolecular segmental interactions in the 3D structure of a protein molecule is considered to be mainly due to hydrophobic interactions and hydrogen bonding. The force that leads to the rupture of hydrogen-bonded structures has been reported from many examples of single-molecule experiments using AFM, but the measurement of the force required to disrupt hydrophobic interaction seems to be difficult since there was almost no report on its measurment.

When compared with the covalent bond, noncovalent bonds are weak in the sense that they can be easily formed and broken at room temperature due to the fluctuation in local thermal energy, whereas covalent bonds are not easily broken once they are formed. The basic building blocks of biological structures are covalently bonded molecules such as DNA, proteins, RNA, carbohydrates, lipids, and ATP, but what gives life to assemblies of these molecules is the noncovalent interactions between covalently built molecules.

Life is thus built on the noncovalent interactions among a considerable number of large and small molecules. Since the noncovalent interactions are relatively weak and are constantly being formed and destroyed, life is characterized as 'dynamic'. For example, DNA is not always in its well-known double-helical structure, but is repeatedly unfolded into a single-stranded form for replication and transcription and folded back to double helix when the work is done. Proteins constantly bind and unbind ligands in cytoplasm as well as on the cell membrane. Cell membrane is made of a large number of phospholipids arranged in a form of 2D leaflets due to hydrophobic interactions among their hydrocarbon 'tails'. Phospholipids form 2D membrane only in water environment. Hydrophobic interactions are completely lost in organic solvents such as chloroform where bilayer structures cannot be formed. Synthetic plastic materials are also built on the noncovalent interaction of covalently built macromolecules but are much harder compared with biological materials, though not as hard as steel or diamond. This is because they are dried or cured and are not immersed in solvent. When immersed in a suitable solvent, they also become soft, but steel or diamond never becomes soft in solvents.

In nano-biomechanics, we are mainly concerned with the mechanical manipulation of noncovalently bonded structures, forming and breaking, for example, hydrogen-bonded double-helical DNA, ligand–receptor complexes on the cell membrane, or creating a hole on the cell membrane to extract cytoplasmic components or to insert plasmid DNA into the cell.

## 5.4 APPLICATION OF EXTERNAL FORCE

To manipulate noncovalently assembled, biological structures by force, we apply a controlled force to a targeted site of the sample molecule or cell. A pulling or pushing force is applied to the sample through the cantilever of AFM or other devices. The laser tweezers are capable of controlling the applicable force with a few pN's, but the nanometer-sized sample must be tagged to a latex sphere of approximately 1 $\mu$m in diameter, which is quite applicable for the manipulation of a long strand of dsDNA, fibrous structures such as microtubules, or large samples such as live cells, but not convenient, for example, for the manipulation of single protein molecule. The force that could be generated by optical trap is less than 100 pN, but now a force larger than 500 pN is available, expanding the range of biological systems that could be studied using this method.

AFM can exert a force over a much wider range by choosing a cantilever whose force constant may be changed from 1 pN/nm and above, for the manipulation a small object such as individual protein molecules. As stated previously, the force involved in biological mechanics is between 1 pN and 1 nN, and AFM is fitted to exert a force over this range. The most typical way of applying force is either to push or pull the sample normal to the substrate surface. It is also possible to apply a nonvertical force, but it is rarely done except in the case of lateral (shear) force measurement.

## 5.5 INTERACTION FORCE BETWEEN MACROMOLECULES

There are several interaction forces to be recognized between macromolecular samples.

### 5.5.1 Exclusion effect

Two random coil polymer molecules, when brought close to each other, do not easily mix their segments. Therefore, when they are grafted on a substrate with increasing surface density, their shape changes from a mushroom to an elongated brush, as shown in Figure 5.3.

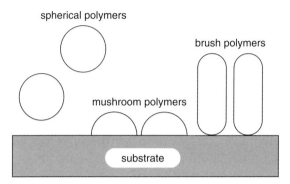

**Figure 5.3** Transition from a polymer mushroom to brush polymer takes place depending on the number density on the solid surface.

Similarly, when the AFM probe and the substrate are coated with randomly coiled polymers, they would repel each other when they are brought into contact and an extra force is required to penetrate the polymer layer on the substrate by the similar layer on the probe.

## 5.5.2 Depletion effect

The coexistence of nonabsorbing polymer in a solution of macromolecules or colloidal particles results in attractive interactions between the macromolecules. It is due to the exclusion of these polymers from around the macromolecules and is called the depletion zone [9, 10]. When two macromolecules are in contact, the depletion zones overlap and the total volume accessible to the polymer increases, which in turn results in an attractive interaction between macromolecules and is called the depletion force. The range of the attraction is directly related to the radius of gyration, whereas the strength is proportional to the osmotic pressure of the polymers.

As we have seen above, the interaction between two macromolecules occurs in several ways, wanted or unwanted. In nanomechanical experiments, it is more difficult to reduce unwanted, nonspecific interactions and single out, hopefully, the specific interaction force than to measure the force itself. In the next few chapters, I will describe experimental results of force measurements at the molecular level, where distinction of 'specific' versus 'nonspecific' interaction is important. The basic tenet is that all interact with all, but in biology, there must be a very finely tuned interaction regime because many of the specific interactions take place over a force range from a few pN to 100 pN, where many nonspecific interactions must be effectively suppressed.

 ## 5.6 WATER AT THE INTERFACE

The importance of water for life has been emphasized repeatedly. Without water, life is impossible, at least on this planet. Water close to the surface of biomolecules and biostructures is all the more important because it might be crucially affecting the activity and/or the faculty of molecular recognition of the protein and DNA. It has been argued that water close to a solid surface is differently structured from the bulk water, although the structure of bulk liquid water itself is not fully understood. Experimental and theoretical works by using computer simulations agree in that the water close to a solid surface has a multilayered structure, where the density of water alternates between lower and higher values compared with that of the bulk water [11, 12].

Higgins et al. reported the presence of structured water layers near solid surfaces using a highly sensitive dynamic AFM technique in conjunction with a carbon nanotube (CNT) probe. They revealed a hydration force with an oscillatory profile that reflects the removal of up to five structured water layers from between the probe and biological membrane surface [13]. They also found that the hydration force can be modified by changing the membrane fluidity by replacing the phospholipids in the membrane. For the experiment such as this one, they used an AFM probe with an extra carbon-nanotube modification as developed by Nakayama and colleagues in their three-step fabrication method [14, 15], (1) purification and alignment of carbon nanotubes by using electrophoresis, (2) transfer of a single aligned nanotube onto a conventional Si tip under the view of a scanning electron microscope, and (3) attachment of the nanotube on the Si tip by carbon deposition, and the result is shown in Figure 5.4.

It is interesting that five layers of structured water were detected by force measurement by using an AFM. The interface phenomenon in these layers is obviously important, and its significance to biological world will be elucidated in the near future. Water layers not only on the lipid bilayer but also around individual protein and DNA molecules may also be studied in a similar manner.

Water is a very important and yet difficult subject, and it is difficult to draw conclusions on the aspect(s) based on which it plays an outstanding role in the biological specificity. One of the sources of controversy is the fact that many phenomena involving water as a component are accompanied by the unusually large

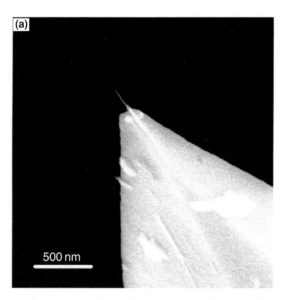

**Figure 5.4** Carbon–nanotube probe was created on the tip an ordinary AFM probe inside a scanning electron microscope. Reproduced from the study by Nakayama [15] with permission.

change in entropy, and there is a plenty of room to discuss the origin of that entropic change. A change in energy is calculated rather directly based on quantum mechanics of chemical structures, but numerical assessment of the entropic change is often model dependent. There are more than several models of water itself and several models on the water at various interfaces, all trying to explain the entropic change accompanying experimentally observed dynamic events.

# Bibliography

[1] Israelachvili, J. N. (1992), 'Intermolecular and Surface Forces', Academic Press, London, Chapter 6.

[2] Ibid., Chapter 12.

[3] Ibid., Chapter 11.

[4] Jeffery, G. A. (1997), 'An Introduction to Hydrogen Bonding', Oxford University Press. Oxford, UK.

[5] Tanford, C. (1980), 'The Hydrophobic Effect', Wiley, New York.

[6] Kauzmann, W. (1959), Some factors in the interpretation of protein denaturation, Adv. Protein Chem., 14, 1-63.

[7] Yoon, R. H. and Ravishankar, S. A. (1996), Long-range hydrophobic forces between mica surfaces in dodecylammonium chloride solutions in the presence of dodecanol, J. Colloid Interface Sci., 179, 391–402.

[8] Craig, V. S. J., Ninham, B. W., and Pashley, R. M. (1998), Study of the long-range hydrophobic attraction in concentrated salt solutions and its implications for electrostatic models, Langmuir, 14, 3326–3332.

[9] Asakura, S. and Oosawa, F. (1954), On the interaction between two bodies immersed in a solution of macromolecules, J. Chem. Phys., 22, 1255–1256.

[10] Asakura, S. and Oosawa, F. (1958), Interaction between particles suspended in solutions of macromolecules, J. Polym. Sci., 33, 183–192.

[11] Bhide, S. Y. and Berkowitz, M. L., (2006), The behavior of reorientational correlation functions of water at the water–lipid bilayer interface, J. Chem. Phys., 125, 094713.

[12] Bhide, S. Y., Zhang, Z., and Berkowitz, M. L. (2007), Molecular dynamics simulations of SOPS and sphingomyelin bilayers containing cholesterol, *Biophys. J.*, 92, 1284–1295.

[13] Higgins, M. J., Polcik, M., Fukuma, T., Sader, J., Nakayama, Y., and Jarvis, S. (2006), Structured water layers adjacent to biological membranes, *Biophys. J.*, 91, 2532–2542.

[14] Nishijima, H., Kamo, S., Akita, S., Nakayama, Y., Hohmura, K. I., Shige, H. et al. (1999), Carbon-nanotube tips for scanning probe microscopy: preparation by a controlled process and observation of deoxyribonucleic acid, *Appl. Phys. Lett.*, 74, 4061–4063.

[15] Nakayama, Y. (2002), Scanning probe microscopy installed with nanotube probes and nanotube tweezers, *Ultramicroscopy*, 91, 49–56.

## CHAPTER SIX

# SINGLE-MOLECULAR INTERACTION FORCES

## Contents

One major purpose of using AFM for force measurement is to quantify the strength of ligand–receptor interaction forces under physiological conditions. Although all interactions may be regarded as ligand–receptor interactions, we classify macromolecular interactions into the following types and describe examples from the literature accordingly.

- Ligand–receptor interactions

- Lectin–sugar interactions

- Antigen–antibody interactions

- GroEL–substrate interactions

- Lipid–protein interactions
- Force anchoring proteins to the membrane
- Receptor mapping
- Protein unanchoring and identification
- Membrane breaking

## 6.1 LIGAND–RECEPTOR INTERACTIONS

### 6.1.1 Biotin–avidin interaction

Florin et al. reported the first study on the interaction force between biotin and avidin pair, which is known as the most stable noncovalent complex in biochemistry [1]. In biochemical work, biotin–avidin pair formation is widely used as the anchoring platform of specifically labeled proteins, DNA, and other molecules and structures. The binding constant of the pair is quoted to be as high as $10^{14-15}$ $M^{-1}$ and is several orders higher than that of other types of noncovalent pair formation, including antigen–antibody, sugar–lectin, enzyme–inhibitor, etc. Hirudin–thrombin complex formation has been reported to be almost as stable as biotin–avidin pair [2].

Biotin is a small molecule essential as the prosthetic group of a certain kind of enzymes involved in the fixation of carbon dioxide, *i.e.*, carboxylases. Their covalent structure is given in Figure 6.1 together with that of avidin. Avidin is a tetrameric protein molecule with a molecular weight of 68,000 found in the white of hen egg. Why in the egg white? It is explained that the presence of avidin stops the proliferation of invading bacteria in the egg by binding biotin, which is vital for the life of bacteria.

Florin et al. used agarose beads, which were coated with biotin molecules by the covalent crosslinker and then exposed to avidin to form biotin–avidin pairs on the bead surface, and the bead was immobilized to a solid substrate. They coated the AFM probe with biotin and approached the agarose bead using

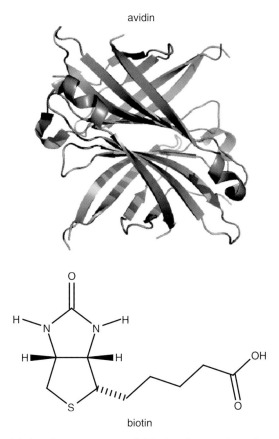

**Figure 6.1** Molecular structure of biotin (bottom) and avidin (top). Biotin binds to the pocket formed on the suface of avidin with an extremely high affinity.

the force mode of AFM in liquid cell. The resulting force curves are reproduced in Figure 6.2. When the number density of active avidin was high, the unbinding force measured as the rupture force was large, but as a substantial fraction of avidin was inactivated by adding free biotin, the rupture force became small and finally the force of ~150 pN was determined as the unbinding force of single pair of avidin and biotin complex. Later, after the realization of loading-force dependence, the force range was reported as between 5 pN and 170 pN when the loading rate was changed [3, 4].

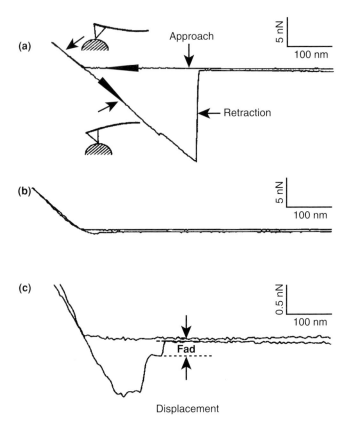

**Figure 6.2** Unbinding event of biotin–avidin pair under an applied tensile force is captured by AFM. (A) A large number of avidin–biotin pairs were unbound with a large force. (B) Most of biotin on the substrate was blocked by adding free avidin; thus the unbinding force was drastically reduced. (C) A blow-up of the ending part of the force curve in (B) revealed a small rupture force corresponding to unbinding of a few bonds including cases where a single bond was ruptured. Reproduced with permission of SCIENCE [1].

Sekiguchi and Ikai compared the results of several groups on the loading rate dependence of the unbinding force of biotin–avidin pair [5] as shown in Figure 6.3. The variation of data in vertical scale is rather large, but most of the data show two regions of linear dependence of the force on the loading rate, suggesting that the energy landscape of the reaction is characterized by double (or multiple) minima.

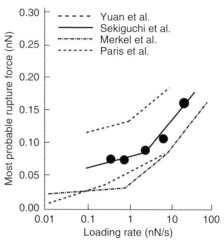

**Figure 6.3** Loading-rate dependence of rupture force of biotin–avidin complex by several different measurements. There are still a wide range of discrepancies among measurements of different groups resulting in almost one order of magnitude difference in the value of activation distance. Reproduced from the study by Sekiguchi and Ikai [5] with permission.

It was surprising that the unbinding force of the most stable noncovalent ligand–receptor pair of biotin–avidin was not much different from that of antigen–antibody pairs or sugar–lectin pairs whose binding constant is in the range of $10^{6-7} M^{-1}$ compared with a much higher value of $10^{14-15} M^{-1}$ for biotin–avidin pairs.

## 6.1.2 Interaction of synaptic-vesicle fusion proteins

Yersin et al. [6] investigated the interactions between synaptic-vesicle fusion proteins using AFM. As shown in Figure 6.4, there are at least three prominent proteins involved in the process of synaptic-vesicle fusion, namely, VAMP 2 on the synaptic vesicle, syntaxin, and SNAP-25 on the target membrane for fusion.

Yersin et al. measured the unbinding force of each pair of the three proteins and that of the remaining protein, and the result is summarized in Figure 6.4. The specific interaction force ranges

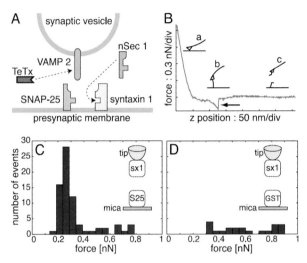

**Figure 6.4** Model illustration of synaptic interaction. The names of participating proteins used in the text are schematically explained here. Reproduced from [6] with permission.

from approximately $100\,\mathrm{pN}$ (VAMP 2 versus syntaxin) to $250\,\mathrm{pN}$ (SNAP-25 versus syntaxin).

## 6.1.3 Interaction between transferrin and its membrane receptor

Yersin also measured the interaction force between transferrin and its receptor either in isolated condition on a mica surface or directly on the live cell surface [7]. Transferrin is an iron ($Fe^{3+}$)-transporting protein in the blood stream and delivers irons to peripheral cells through binding to the membrane-bound receptors. Once bound to the receptor, transferrin is internalized by way of endocytosis mechanism and with the lowering of pH in the endosome, ferric iron is released from transferrin and reduced to $Fe^{2+}$ to be transported out of the endosomal vesicle. The remaining complex of transferrin and its receptor are recycled to the cell membrane where apo-transferrin is released to the blood stream for reuse. Thus, the pH-dependent interaction force between holo- and apo-transferrin with the receptor was clearly distinguished. The unbinding of transferrin from the cell-surface receptor gave a rupture force in the same range.

## 6.2 SUGAR–LECTIN INTERACTIONS

Lectins are proteins that have rather strong affinity toward specific sugar moieties. For example, the well-known lectin, concanavalin A (Con A) binds to $\alpha$-D-mannoside (and to a lesser extent with $\alpha$-D-glucoside) with the binding constant of $1-2 \times 10^6$ $M^{-1}$. Con A is extracted from Jack bean seeds, but its intrinsic biological function is not known. It is a dimer or tetramer of the subunit with $MW = 26,000$, having 2 or 4 binding sites per molecule.

The unbinding force of concanavalin A from a specific ligand was measured using AFM. The unbinding force depends on the loading rate and values ranging from $47 \pm 9$ pN with the loading rate of 10 nN/s [8], $96 \pm 55$ pN [9], from 75 to approximately 200 pN [10] have been reported. Another example of lectin unbinding using wheat germ agglutinin is given below [11].

 ## 6.3 ANTIGEN–ANTIBODY INTERACTIONS

Antigen–antibody interaction has been studied extensively by using AFM. One of the first of such experiment was done by Allen et al. [12]. They reported that the unbinding events between the ferritin-coated AFM probe and antiferritin antibody immobilized on a substrate could be quantized with respect to the unbinding force of the smallest step of $49 \pm 10$ pN. Ferritin is a large molecule composed of 24 subunits containing approximately 4500 ferric ions ($Fe^{3+}$), and therefore the binding efficiency with its antibody is high.

Hinterdorfer and colleagues focused on the quantitative measurement of unbinding force of antigen–antibody pairs and applied their overall results to develop a new mode of AFM technology, *i.e.*, TREC method of simultaneously imaging antigen molecules and identifying them through the specific interaction with the antibody molecules immobilized on the AFM probe [13, 14]. Antibody molecules are immobilized on the AFM probe via covalent cross-linkers with a long polyethylene glycol spacer. After mounting the modified probe on an AFM, the sample surface having antigen molecules is scanned using the probe in dynamic mode for imaging. Where there are no antigen molecules on the surface, normal imaging is done by monitoring the change in vibration amplitude of the cantilever, and when there is positive interaction between the sample and the antibody on the probe, the upper level of cantilever oscillation becomes limited because of the presence of PEG crosslinker of limited length. By monitoring the change in both the change in total amplitude as well as the limit of the upper level of oscillation, the TREC mode images and identifies specific antigen in a single scan.

Red blood cell imaging and blood-type specific interaction force were reported [15]. In addition to the topographical imaging, they reported the result of measurement of the unbinding force between antibody (anti-A) and the RBC antigen A.

## 6.4 GroEL and Unfolded-Protein Interactions

Interaction between chaperonine and denatured protein was studied using force spectroscopy [16]. Since it is important to have a denatured protein interacting with active GroEL, Sekiguchi et al. modified the AFM probe with pepsin, which has an unfolded conformation at neutral pH. In order to minimize the possibility of denaturing GroEL by pushing it too strongly with the AFM probe, they repeated approach and retraction cycle of the cantilever from the position approximately 1 $\mu$m above the sample and obtained force curves, which did not show upward deflection of the cantilever in the approach regime, whereas showed only downward deflection on the retraction regime. Pepsin was immobilized on the probe using a crosslinker having a relatively long spacer. 'Noncompressive' force curves were thus obtained, which promised an intact condition of GroEL under the AFM probe. The $F$–$E$ curves thus obtained showed a plateau force of approximately 45 pN with a plateau extension of approximately 12 nm in the absence of ATP. The results are summarized in Figure 6.5

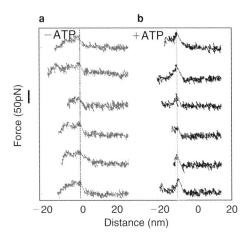

**Figure 6.5**   Unbinding force curves of denatured pepsin from GroEL by using noncompressing technique in AFM: (a) in the absence of ATP; (b) in the presence of ATP. The peak force values were nearly the same in (a) and (b), but the force plateau was only observed in (a). Reproduced from the study by Sekiguchi [16] with permission.

When ATP was added at a concentration of 50 mM, the plateau force was replaced with a force peak, again with a height value of 45 pN. The result indicated that, in the absence of ATP, seven subunits of GroEL had a closed configuration with seven binding sites lining the edge of the ring structure with a small diameter so that the denatured pepsin on the probe had an easy access to most or all of the binding sites. Pulling up unbound pepsin from one binding site after another gave rise to a plateau force of approximately constant force. If the resolution is better, it would be possible to distinguish multiple force peaks corresponding to individual unbinding events. In the presence of ATP, on the contrary, the binding sites on GroEL were aligned on a ring of larger diameter due to the conformational change of each subunit, thus making it difficult for the denatured pepsin on the probe to span multiple binding sites.

The noncompressive method introduced in this study is a good solution to be pursued in similar studies in the future to avoid destruction of the sample proteins under AFM probes.

Very recently, the ATP-dependent conformational change of GroEL has been visualized by the fast-scanning AFM by Yokokawa et al. [17, 18]. The fast-scanning AFM developed by Ando et al. has been a significant development in the instrumentation of AFM in recent years [19]. GroEL is known to undergo conformational changes and induces GroES binding in the ATP-dependent reaction cycle. Using the real-time 3D observation system, they visualized the GroES binding to and dissociating from individual GroEL with a lifetime of 6 s ($k = 0.17$ s$^{-1}$). They also observed ATP/ADP-induced open–closed conformational changes of individual GroEL in the absence of GroES and substrate proteins. These results indicated that GroEL has an ATP-bound prehydrolysis open form and an ADP-bound closed form. The ATP hydrolysis in open form destabilizes its open conformation and induces the 'from open to closed' conformational change of GroEL.

## 6.5 LIPID–PROTEIN INTERACTIONS

Proteins with a high affinity to lipids are found in serum as lipoproteins and in biomembranes as intrinsic membrane proteins. Little mechanical work has been done on lipoproteins, but quite a few articles have been published on the measurement of the anchoring force of membrane proteins to the lipid bilayer.

**GPI-anchored protein:** Some proteins with an affinity toward the lipid membrane is posttranslationally modified with fatty acids or phospholipids at a certain position on their primary structure. Glycosylphosphatidylinositol (GPI)–modified proteins are composed of a hydrophobic phosphatidylinositol group linked through a carbohydrate-containing linker (glucosamine and mannose linked to phosphorylethanolamine residue) to the C-terminal amino acid of the mature protein. The two fatty acids within the hydrophobic phosphatidylinositol group anchor the protein to the membrane. Treatment with phospholipase C releases GPI-linked proteins from the outer cell membrane. The T-cell marker Thy-1, acetylcholinesterase, as well as both intestinal and placental alkaline phosphatase are known to be GPI linked. GPI-linked proteins are thought to be preferentially localized in lipid rafts, suggesting a high level of organization within microdomains in the plasma membrane. Extraction of GPI-anchored proteins from the lipid membrane has been reported by Cross et al. [20]. They studied the adhesion probability and the adhesion force of GPI-modified alkaline phosphatase to the lipid bilayer. They provided proteins with and without GPI anchor and compared the interaction mode of the two proteins and found that the GPI anchors increased the adhesion frequency significantly. An adhesion force of $350 \pm 200\,pN$ is measured between GPI-anchored alkaline phosphatase and supported phospholipid bilayers of dipalmitoylphosphatidyl-choline (DPPC) presenting structural defects such as holes [21]. In the absence of defects, the adhesion force became smaller ($103 \pm 17\,pN$) and the adhesion frequency was reduced. Their results suggested that the GPI-anchored alkaline phosphatase inter-acted strongly with the edge of the hole but less so with an intact holeless membrane, suggesting that the GPI anchor cannot easily penetrate into the lipid bilayer. Some membrane proteins are mod-ified with palmitoyl or myristoyl fatty acid moieties at the cysteine residues at specific sites in the primary structure. Desmeules et al. reported the interaction between recoverin, a calcium–myristoyl switch protein, and lipid bilayers using force spectroscopy [22]. An adhesion force of $48 \pm 5\,pN$ was measured between recov-erin and supported phospholipid bilayers in the presence of $Ca^{2+}$. No binding was observed in the absence of $Ca^{2+}$ nor with non-myristoylated recoverin. Their results are consistent with previously measured extraction forces of lipids from membranes. Lipid-protein interaction *per se* has not been studied on a one-to-one basis. The measurment of the force to extract fatty acids from the binding pockets in serum albumin forms an interesting subject for the future.

## 6.6 ANCHORING FORCE OF PROTEINS TO THE MEMBRANE

Bell, in his seminal paper in 1978 [23], estimated the anchoring force of glycophorin A to the red cell membrane to be in the range of $100\,pN$ in the absence of specifically bound lipids, or $260\,pN$

$(2.6 \times 10^{-5}$ dyne) in case of a surrounding layer of boundary lipids. His estimate was based on the free energy of hydrophobicity of the membrane-spanning segment of the protein and an estimated value of activation length of approximately 2–3 nm. Experimentally, Evans et al. reported the result obtained using membrane force probe (BFP), which stated that the force required to extract glycophorin A from the red blood cell membrane was 26 pN $(2.6 \times 10^{-6}$ dyn) [24], a significantly smaller force compared with the prediction given by Bell. They used a monoclonal antibody raised against glycophorin A and confirmed the transfer of fluorescence label on glycophorin A from the blood cell to the probe. They also used the specific lectin against glycolipids specific to blood type A cell and obtained a similar value for the glycolipid extraction.

Afrin and Ikai reported the result of pulling experiment of glycophorin A using a probe modified with WGA, which has a specific affinity toward the sialic acid and N-acetyl-D-glucosamine residues of the sugar moiety of the protein [25]. Their result showed that glycophorin A could not be extracted with a force of at least 70 pN. With this force, glycophorin A was pulled out of the membrane up to 1–3 $\mu$m, most likely with the formation of a lipid tether trailing behind it. After the pullout, the force plateau was terminated with a step force of 70 pN. The final rupture event is either unbinding of lectin–glycophorin A interaction or extraction of glycophorin A from the lipid membrane including the cutting of the lipid tether. Thus, it is not possible to extract glycophorin A with a force less than 70 pN.

The pulling speed (loading rate) of course, should be considered when the two force values are compared.

 ## 6.7 RECEPTOR MAPPING

The unbinding force between ligand and receptor or antigen and antibody can be used as a marker to probe for the presence of particular proteins or other molecules on the cell surface. An early example of such experiment was conducted by Gad and Ikai [26] on the yeast surface. They first showed that rigid and spherical yeast cells were difficult to be imaged using AFM; however, when cells were embedded in a thin layer of agarose, yeast cells were stably imaged and kept alive for a long duration of time [10]. They then used a modified AFM probe with concanavalin A and repeated contact and retraction cycle to the cell surface while shifting the lateral position systematically using the force volume mode of the AFM. In one cycle of approach and retraction, the force curve showed an entrapment of tensile molecules between the probe and the cell surface and the force curve. It was concluded that they were mannan molecules known to be present on the yeast cell surface. As the probe is retracted from the cell surface, mannan molecules were sequentially detached from the probe with a rupture force ranging between 70 and 200 pN. By using this high level of interaction force, several maps were presented, which showed a nonuniform distribution of mannan molecules (Figure 6.6).

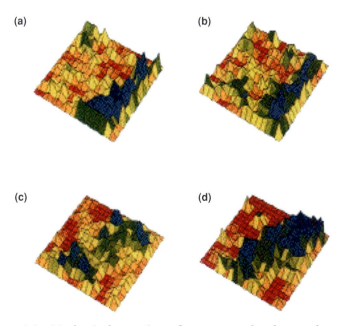

**Figure 6.6**  Mechanical mapping of mannan molecules on the surface of live yeast cells by using an AFM probe coated with the lectin, concanavalin A. Reproduced from the study by Gad and Ikai [26] with permission.

Kim et al. [27] mapped the distribution of vitronectin receptors on the surface of live fibroblast cells. They used a colloidal AFM probe of diameter of approximately 20 $\mu$m, which was modified with vitronectin molecules, and measured the de-adhesion force at 64 points on the cell surface. The colloidal probe was chosen so as to increase the contact area with the cell for the purpose of covering a wider area than in the case where the original AFM probe was used. By using a colloidal probe, they traded in the possibility of mapping at a single-molecule resolution for mapping a wider area within a short time. They introduced an integrated quantity of a part of the force curve showing adhesive interactions, calling it 'unbinding work'. The result of the study was thus not of single molecule resolution, but the study fulfilled the function of mapping the dominant presence of receptors to vitronectin, *i.e.*, integrins. The result was in good agreement with that of fluorescence staining of the receptors with a specific antibody. Slight difference between the results of the two methods may be due to the fact that the AFM-based method maps the presence of the receptors on the dorsal side of the cell, whereas the fluorescence method is likely to map the presence on the ventral as well as dorsal sides of the cell. The advantage of the AFM-based mapping method lies in the fact that the cell can be kept alive while and after mapping, so that mapping of the same or different receptors on exactly the same cell can be done at different times.

One caveat of mapping of receptors on the cell surface is that the unbinding force required of ligand–receptor or antibody–antigen pairs must be significantly less than the force required to extract the receptors from the cell membrane. Otherwise, in every mapping process, the target cell loses a certain number of receptors from the scanned part of its surface, which may result in abnormal cellular physiology. Extension of membrane tubes as tethers should also be suppressed to obtain good results. The mechanism and the force range of tether formation are needed for further advancement of cell-surface mapping of receptors. The membrane phospholipids have a tendency to be pulled out of the cell membrane as a thin tube following the membrane proteins on the AFM probe, creating an extremely complicated surface phenomena. To avoid protein extraction and the lipid tether formation, the unbinding of ligand–receptor pair should take place at a force less than 50 pN. This receptor unanchoring possibility was recognized by Ikai and associates at the early stage of receptor mapping [28].

## 6.8 PROTEIN UNANCHORING AND IDENTIFICATION

Membrane proteins are classified into two groups, the extrinsic and the intrinsic membrane proteins. Extrinsic proteins are electrostatically adsorbed on the phospholipid membrane and can be easily washed out by the addition of chelating agents, which weaken the electrostatic force mediated by metal ions. We are here concerned with intrinsic proteins, which have hydrophobic segments that span the hydrophobic interior of the lipid bilayer membrane. Some of them have a single segment and others have two or even more than ten membrane-spanning segments and are firmly anchored to the lipid membrane. But how firmly?

Direct measurement of the thermodynamic affinity of the intrinsic membrane protein to the lipid bilayer membrane is difficult because the solubility of the intrinsic membrane proteins is quite small. However, it is possible, for example, to measure the force to extract such proteins from the membrane by the application of a tensile force by using AFM. Several attempts have been reported in this regard. Among them, the work of Afrin et al. [28, 29] is the most complete. Instead of using antibodies against particular membrane protein to pull them out of the membrane, Afrin et al. used bifunctional amino-reactive covalent cross-linkers to form strong bonds between the AFM probe and membrane proteins on the surface of fibroblast cells. After formation of such bonds, the AFM probe was moved away from the cell surface together with membrane proteins that were tethered to the probe with the cross-linker. The experimental setup is given in Figure 6.7.

They reported that the final rupture force was no less than 450 pN. The loading-rate dependence was very low. The presence of extracted proteins on the probe was verified from the force

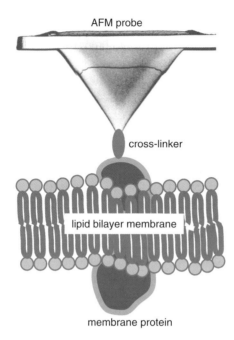

**Figure 6.7**  Experimental setup to extract membrane proteins by using modified AFM probe with covalent cross-linkers.

curve measurement using a used probe on the cell surface on a freshly prepared silicon surface modified with cross-linkers. It was argued that the final rupture force might have been the force to cut a lipid tether that trailed behind the membrane–protein. In that case, membrane–protein interaction is even stronger.

It was most likely that the majority of the extracted proteins were integrins, which are usually linked to the cytoskeleton. The links between integrins, linker proteins (mainly talin), and the cytoskeletal proteins should have been severed with a force equal to or less than 450 pN.

 **6.9 MEMBRANE BREAKING**

In the preceding sections, we have seen the ligand unbinding and/or protein uprooting events on the cell surface. What will happen if the AFM probe is pushed onto the cell membrane a little more strongly? This question leads to the mechanical penetration of a cell membrane with a nano-probe. Butt et al. and Künneke et al. reported observation of characteristic force curves when the probe was pushed onto a LB film formed on a solid surface [30–32]. The dip in the approach part of the force curve has been identified as representing the failure point of the lipid membrane as shown in Figure 6.8 [32].

In the case of LB film failure, the tethering state of lipid molecules to the solid surface is an important parameter. Naturally, when the lipid molecules are tethered to the solid surface, they would be pushed down by the incoming probe, but to be moved aside is difficult. Therefore, a tethered membrane would require a larger force than an untethered film for mechanical failure.

On the live cell surface, it is possible to obtain similar dips when the AFM probe is strongly pushed on to the cell surface, but penetration on the live cell membrane is a little more

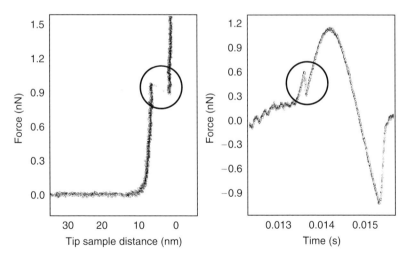

**Figure 6.8**  Membrane penetration event as observed in AFM experiment. Force curves as obtained by conventional force microscopy (left) and by pulsed force microscopy (right) on POPS bilayers systems. Both graphs exhibit breakthrough events in the contact regime. Reproduced from the study by Künneke [32] with permission.

complex due to the lack of reliable verification of the penetration event. Uehara et al. tried to correlate the observed dip with the recovery of mRNA from the cytoplasm as a clear evidence of membrane penetration, but the correlation between the two events was not significant enough to establish a close relationship between them [33].

## Bibliography

[1] Florin, E. L., Moy, V. T., and Gaub, H. E. (1994), Adhesion forces between individual ligand–receptor pairs, *Science*, 264, 415–417.

[2] Chang, J. Y. (1989), The hirudin-binding site of human $\alpha$-thrombin identification of lysyl residues which participate in the combining site of hirudin–thrombin complex, *J. Biol. Chem.*, 264, 7141–7146.

[3] Merkel, R., Nassoy, P., Leung, A., Ritchie, K., and Evans, E. (1999), Energy landscapes of receptor–ligand bonds explored with dynamic force spectroscopy, *Nature*, 397, 50–53.

[4] Sekiguchi, H., Ikai, A., Arakawa, H., and Sugiyama, H. (2006), AFM analysis of interaction forces between bio-molecules using ligand-functionalized polymers, *e-Journal of Surface Science and Nano Technology*, Published by Surface Science Society of Japan, 4, 149–154.

[5] Sekiguchi, H., and Ikai, A. (2006), A method of measurement of interaction force between ligands and biological macro-molecules (in Japanese), *Hyoumenkagaku*, 27, 436–441.

[6] Yersin, A., Hirling, H., Steiner, P., Magnin, S., Regazzi, R., Huni, B. et al. (2003), Interactions between synaptic vesicle fusion proteins explored by atomic force microscopy, *Proc. Natl. Acad. Sci. USA*, 100, 8736–8741.

[7] Yersin, A., and Ikai, A. (2007), Exploring transferrin-receptor interactions at the single molecular level, *Biophys. J.* (in press).

[8] Ratto, T. V, Langry, K. C., Rudd, R. E., Balhorn, R. L., Allen, M. J., and McElfresh, M. W. (2004), Force spectroscopy of the double-tethered concanavalin-a mannose bond, *Biophys. J.*, 86, 2430–2437.

[9] Touhami, A., Hoffmann, B., Vasella, A., Denis, F. A., and Dufrene, Y. F. (2003), Probing specific lectin–carbohydrate

interactions using atomic force microscopy imaging and force measurements, *Langmuir*, 19, 1745–1751.

[10] Gad, M., Itoh, A., and Ikai, A. (1997), Mapping cell wall polysaccharides of living microbial cells using atomic force microscopy, *Cell Biol. Int.*, 21, 697–706.

[11] Krotkiewska, B., Pasek, M., and Krotkiewski, H. (2002), Interaction of glycophorin A with lectins as measured by surface plasmon resonance (SPR), *Acta Biochim. Pol.*, 49, 481–490.

[12] Allen, S., Chen, X., Davies, J., Davies, M. C., Dawkes, A. C., Edwards, J. C. et al. (1997), Detection of antigen–antibody binding events with the atomic force microscope, *Biochem.*, 36, 7457–7463.

[13] Stroh, C. M., Ebner, A., Geretschlager, M., Freudenthaler, G., Kienberger, F., Kamruzzahan, A. S. et al. (2004), Simultaneous topography and recognition imaging using force microscopy, *Biophys. J.*, 87, 1981–1990.

[14] Kienberger, F., Ebner, A., Gruber, H. J., and Hinterdorfer, P. (2006), Molecular recognition imaging and force spectroscopy of single biomolecules, *Acc. Chem. Res.*, 39, 29–36.

[15] Touhami, A., Othmane, A., Ouerghi, O., Ouada, H. B., Fretigny, C., and Jaffrezic-Renault, N. (2002), Red blood cells imaging and antigen-antibody interaction measurement, *Biomol.Eng.*, 19, 189–193.

[16] Sekiguchi, H., Arakawa, H., Taguchi, H., Ito, T., Kokawa, R., and Ikai, A. (2003), Specific interaction between GroEL and denatured protein measured by compression-free force spectroscopy, *Biophys. J.*, 85, 484–490.

[17] Yokokawa, M., Wada. C., Ando, T., Sakai, N., Yagi, A., Yoshimura, S. H. (2006), Fast-scanning atomic force microscopy reveals the ATP/ADP-dependent conformational changes of GroEL, *EMBO J.*, 25, 4567–4576.

[18] Yokokawa, M., Yoshimura, S. H., Naito, Y., Ando, T., Yagi, A., Sakai, N. et al. (2006), Fast-scanning atomic force microscopy reveals the molecular mechanism of DNA cleavage by ApaI endonuclease, *IEE Proc. Nanobiotechnol.*, 153, 60–66.

[19] Ando, T., Kodera, N., Takai, E., Maruyama, D., Saito, K., and Toda, A. (2001), A high-speed atomic force microscope for studying biological macromolecules, *Proc. Natl. Acad. Sci. USA*, 98, 12468–12472.

[20] Cross, B., Ronzon, F., Roux, B., and Rieu, J. P. (2005), Measurement of the anchorage force between GPI-anchored alkaline phosphatase and supported membranes by AFM force spectroscopy, *Langmuir*, 21, 5149–5153.

[21] Rieu, J. P., Ronzon, F., Place, C., Dekkiche, F., Cross, B., and Roux, B. (2004), Insertion of GPI-anchored alkaline phosphatase into supported membranes: a combined AFM and fluorescence microscopy study, *Acta Biochim. Pol.*, 51, 189–197.

[22] Desmeules, P., Grandbois, M., Bondarenko, V. A., Yamazaki, A., and Salesse, C. (2002), Measurement of membrane binding between recoverin, a calcium-myristoyl switch protein, and lipid bilayers by AFM-based force spectroscopy, *Biophys. J.*, 82, 3343–3350.

[23] Bell, G. I. (1978), Models for the specific adhesion of cells to cells, *Science*, 200, 618–627.

[24] Evans, E., Berk, D., and Leung, A. (1991), Detachment of agglutinin-bonded red blood cells. I. Forces to rupture molecular-point attachments, *Biophys. J.*, 59, 838–848.

[25] Afrin, R. and Ikai, A. (2006), Force profiles of protein pulling with or without cytoskeletal links studied by AFM, *Biochem. Biophys. Res. Commun.*, 348, 238–244.

[26] Gad, M. and Ikai, A. (1995), Method for immobilizing microbial cells on gel surface for dynamic AFM studies, *Biophys. J.*, 69, 2226–2233.

[27] Kim, H., Arakawa, H., Osada, T., and Ikai, A. (2003), Quantification of cell adhesion force with AFM: distribution of vitronectin receptors on a living MC3T3-E1 cell, *Ultramicroscopy*, 97, 359–363.

[28] Afrin, R., Arakawa, H., Osada, T., and Ikai, A. (2003), Extraction of membrane proteins from a living cell surface using the atomic force microscope and covalent crosslinkers, *Cell Biochem. Biophys.*, 39, 101–117.

[29] Afrin, R., Yamada, T., and Ikai, A. (2004), Analysis of force curves obtained on the live cell membrane using chemically modified AFM probes, *Ultramicroscopy*, 100, 187–195.

[30] Butt, H. J. and Franz, V. (2002), Rupture of molecular thin films observed in atomic force microscopy. I. Theory, *Phys. Rev. E.*, 66, 031601–1031601–9.

[31] Franz, V., Loi, S., Müller, H., Bamberg, E., and Butt, H. J. (2002), Tip, penetration through lipid bilayers in atomic force microscopy, *Colloids Surf. B. Biointerf.*, 23, 191–200.

[32] Künneke, S., Krüger, D., and Janshoff, A. (2004), Scrutiny of the failure of lipid membranes as a function of head-groups, chain length, and lamellarity measured by scanning force microscopy, *Biophys. J.*, 86, 1545–1553.

[33] Uehara, H. (2007) Detection of mRNA in single cells using AFM nanoprobes. Ph.D. thesis. *Tokyo Inst. of Technology.*

## CHAPTER SEVEN

# SINGLE-MOLECULE DNA AND RNA MECHANICS

## Contents

## 7.1 STRETCHING OF DOUBLE-STRANDED DNA

Double-stranded DNA (dsDNA) is a rather stiff molecule having a persistence length of 50–60 nm compared with less than 1 nm for a single-stranded DNA (ssDNA) or RNA. It means that a short dsDNA behaves like a rod rather than a flexible string. But since dsDNA is often thousand times longer than its persistence length, it has all the features of a randomly coiled polymer as a whole. It coils up into a roughly spherical shape but is far more expanded compared with ssDNA of a similar contour length. If positively charged molecules such as polyamines are added, coiled DNA becomes dramatically compacted. Stretching of dsDNA from a coiled state as well as from a compacted state is, therefore, an interesting experiment; first to check the validity of the result of polymer theory concerning the elasticity of randomly coiled chains, and second to observe the change of the stretching curve upon interaction with polyamines. Bustamante and colleagues showed the mechanics of dsDNA stretching for the first time using the laser trap method [1]. They immobilized one end of dsDNA of a few $\mu$m long to a latex bead of a few $\mu$m in diameter and immobilized the other end to a non-moving wall. They then applied a laser trap force to the latex bead and pulled it away from the wall with dsDNA trailing behind the bead. The resulting $F-E$ curve is shown in Figure 7.1.

The DNA was stretched close to its full contour length, which is approximately equal to 0.34 nm × (number of base pairs, bp), with an increase of tensile force from 0 to approximately 80 pN. Then, the force, instead of increasing to the covalent-bond-breaking value of a few nanonewtons, remained at a constant value of approximately 65 pN until the extension reached about twice the contour length, and finally the force attained the bond-breaking value. The constant stretching force of 65 pN is a novel finding, and the result showed that the original B-form dsDNA is stretched to a new form, which is stable under an applied tension. This form is termed as S-form and has base pairs at slanted angles

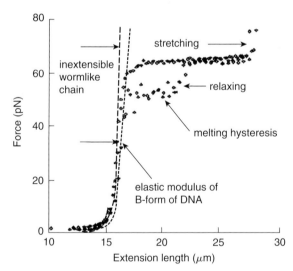

**Figure 7.1** dsDNA was stretched from the two ends by laser tweezers. Stretching up to 15 $\mu$m corresponds to entropic stretching of the B-form dsDNA, and after a sharp increase in force, B-form resulted in a new form called S-DNA. Reproduced from the study by Smith et al. [1] with permission of SCIENCE.

against the helix axis. Whether the S-form DNA has a biologically meaningful role has to be analyzed in the future, but at least we learned that over-stretching of dsDNA beyond the contour length of B-form leads to a new, semi-stable conformation, and this fact must be remembered in the application of DNA in production processes of artificial device.

In a sense, over-stretching of DNA is a phenomenon that is analogous to the necking of plastic materials after being stretched beyond the yield point. As dislocations and bond breaking are taking place in the necking region of plastic fracture, an extensive rearrangement of noncovalent bonding structure must be occurring in DNA.

When proteins are bound to dsDNA, the stretching curve showed multiple force peaks corresponding to forced unbinding of the proteins one after another. It is most interesting to observe the force curves obtained in the process of chromosomal-DNA stretching, where uncoiling chromosome and unbinding of chromosomal proteins were recorded as multiple force peaks self-arranged as a saw tooth pattern [2, 3]. The rupture strength of DNA–protein complex structure was measured to be in the range of 100–200 pN [4].

The $F-E$ curve of dsDNA stretching was fitted to the theoretical $F-E$ curve of a randomly coiled polymer chain. The original $F-E$ curve is expressed by using a modified Langevin function, where $p$ is the persistence length rather than the segmental length.

$$\frac{L}{L_0} = \left[ \coth\left( \frac{Fp}{k_B T} \right) - \frac{k_B T}{Fp} \right] \tag{7.1}$$

Smith et al. presented an interpolation formula as given below [1].

$$F = \frac{k_B T}{p} \left[ \frac{1}{4}\left( 1 - \frac{L}{L_0} \right)^{-2} - \frac{1}{4} + \frac{L}{L_0} \right] \tag{7.2}$$

where $p$, $L$, and $L_0$ are, respectively, the persistence length, extended length, and the contour length of ds DNA. This equation has since been widely used as a convenient fitting equation to the

$F - E$ curves of unstructured polymer chains by changing $p$ as an adjustable parameter. The value of $p$ that gives the best-fit curve to the experimental $F - E$ curves is taken as the persistence length of the polymer chain under the given experimental conditions [5]. Referring back, the persistence length, $p$, is defined as the characteristic length along the polymer chain over which the directional correlation of the chain decays to $1/e$.

A dsDNA can thus be stretched with relatively small forces, such as an electrophoretic force [6, 7] and the meniscus receding force [8, 9].

 ## 7.2 HYBRIDIZATION AND MECHANICAL FORCE

Unzipping of dsDNA by force is another attractive experiment to determine the tensile strength of hydrogen bonds that stabilize base pairs. Early experiment by Lee et al. used modified substrate and an AFM probe with ssDNA of complementary base sequences [10, 11]. After the formation of dsDNA between the strands on the probe and the substrate, the distance between them was increased, consequently applying a shear force to the dsDNA. Rupture force values that depended on the length of complementary base sequence were obtained as shown in Figure 7.2. Breaking the double-stranded structure of DNA by shearing required larger forces compared with unzipping hair-pin DNA because, in the latter case, hydrogen bonds are ruptured one by one in a sequential manner, whereas in the former case, multiple hydrogen bonds arranged in series must be ruptured simultaneously in a cooperative manner.

DNA hybridization is an important experimental tool in molecular biology and is now frequently used in nanotechnology as a tool to align nanotubes and other nanomaterials according to designed patterns on a solid surface. A pattern is created on the surface by using a ssDNA with known base sequences, and the members to be placed at specific sites on the pattern are tagged with DNA with complementary base sequences to the one on

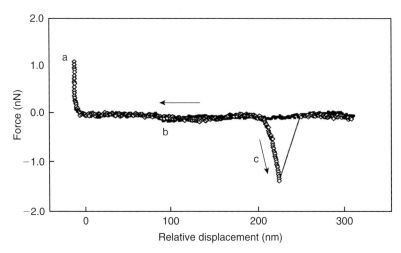

**Figure 7.2** Complementary DNA strands were used to form dsDNA between the AFM probe and the solid substrate and then shear zippered, giving force curves as shown in the figure. Reproduced from the study by Lee et al. [10] with permission.

the patterned surface [12]. Then, in solution, they are mixed and allowed to form a designed distribution of nanomembers according to the rules of hybridization of DNA strands. When DNA is used as a guide for the patterning but not a part of the final structure, it is destroyed at high temperature. DNA mechanical device was presented [13].

## 7.3 CHAIN DYNAMICS AND TRANSITION OF DNA AND RNA

As mentioned above, DNA is known to form especially tight coils when polyamine with a high density of positive charge is added [14]. It is quite natural to evoke electrostatic interactions between negatively charged DNA and positively charged additives to explain this observation. The most interesting aspect of this phenomenon is that the change from relatively expanded randomly coiled state of dsDNA to a condensed state with polyamine occurs within a narrow concentration range of the latter, which reminds us of a phase transition at the macroscopic level. This observation has been obviously studied in an effort to relate it to the compaction of genomic DNA in the nucleus.

The volume of the genomic DNA of one human being is approximately 0.1 L and the cumulative length is approximately $3 \times 10^{10}$ km, long enough to cover more than 100 round trips between the Earth and the Sun. Therefore, one can pull out enormously long stretch of DNA from a seemingly inconsequential mass of DNA. Thus, it has been shown that a single string of dsDNA could be continuously pulled out of a small glob of fluorescently labeled compacted DNA condensate [15].

## 7.4 DNA–PROTEIN INTERACTION

An interesting application of the force mode of the AFM is in the identification of force peaks that were observed when nucleosomal DNA was pulled out. If there are any structural elements of measurable rigidity, one can observe force peaks corresponding to the breakdown of each structural unit and the length of each structural unit. An example is given in Figure 7.3. There are a large number of proteins that bind to specific sequences of genomic DNA and influence the expression of particular genes. Such proteins are called transcription factors because they control transcriptional levels of proteins. The transcription factors do not bind directly to the structural genes that they are supposed to control. Instead, they first bind to a specific locus on DNA called cognate regions. There are several ways to identify the binding regions of transcription factors to DNA by using an AFM. First, the protein–DNA complexes can be imaged under AFM to identify the presence of globular protein molecules at fixed positions along an elongated DNA strand.

The rupture force of transcription factors from DNA has been measured by Jiang et al. on specific interaction between ZmDREB1A, a transcription factor from maize, and its DNA

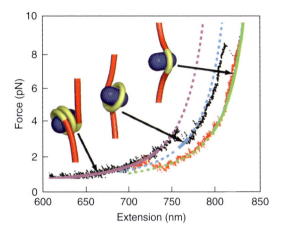

**Figure 7.3** Nucleosomal units were unraveled by the application of tensile force of a few pNs by using optical tweezers. Reproduced from the study by Mihardja et al. [16] with permission.

responsive element, dehydration-responsive element (DRE) with core sequence A/GCCGAC [17]. Single-molecular interaction forces of ZmDREB1A with DRE A/GCCGAC were determined to be $101 \pm 5$ and $108 \pm 3$ pN, respectively.

Yu et al. reported a value of unbinding force of a fragment of another transcription factor, TINY, efficiently bound to DRE with the core sequence A/GCCGAC. The single-molecule forces between TINY and DRE A/GCCGAC measured using atomic force microscopy were $83.5 \pm 3.4$ and $81.4 \pm 4.9$ pN, respectively [18]. They found that either a single-base substitution of the DRE core sequence or a point mutation of the key amino acid in TINY DNA-binding domain considerably reduced the binding strength. Even though they did not measure the loading-rate dependence of the rupture force, relative change due to substitution in base sequence or in amino acid sequence gave biological significance to their measurements when the result was compared with the biological activities of the mutant species.

Whether the use of AFM is a practical means to compare the activity of various mutant species of transcription factors will be analyzed in the future because the AFM-based technology requires some expertise in handling of the instrument as well as in the preparation of the sample in a most appropriate form. In addition, since the values of rupture force for many biologically meaningful interactions measured so far ranged between 10 and 100 pN, a high-precision measurement is strongly required when comparisons are to be made between measurements from different groups. One should also be careful while handling nonspecific interactions. Furthermore, we have very little knowledge of water structure close to the sample surface, which might contribute considerably to the rupture force being measured.

## 7.5 Prospect for Sequence Analysis

If one can stretch DNA molecule as stated above, there appears a prospect of sequencing it using certain imaging technology such as electron microscopy or scanning probe microscopy. For sequencing, it is advantageous to have an ssDNA stretched on a solid surface so that imaging is done to identify four kinds of bases in an exposed state rather than sequestered in the double-helical structure. It is easy, as explained previously, to extend dsDNA and place it on a solid surface, but stretching ssDNA requires a larger force because it has much shorter persistence length than dsDNA, and it is difficult at this moment to maintain it in a stretched form on a solid surface. When ssDNA is placed on a solid surface in a stretched conformation, is it possible to sequence it with required accuracy and speed? The theoretical resolution is high enough for both TEM and AFM for the identification of the four kinds of bases linked to nucleoside backbone, but, in practice, it would be very difficult to achieve the goal. A recent report by using a vacuum spray method to adsorb ssDNA on a solid surface in a stretched conformation shows some promise for the future [19].

Sequencing denatured protein using an SPM has a prospect worth challenging. Again, straightening a coiled chain on a solid surface for imaging is required, but the accuracy is not so demanding because one can use the genome database for searching candidate proteins.

# Bibliography

[1] Smith, S. B., Cui, Y., and Bustamante, C. (1996), Overstretching B-DNA: the elastic response of individual double-stranded and single-stranded DNA molecules, *Science*, 271, 795–799.

[2] Hizume, K., Yoshimura, S. H., Maruyama, H., Kim, J., Wada, H., and Takeyasu, K. (2002), Chromatin reconstitution: development of a salt-dialysis method monitored by nano-technology, *Arch. Histol. Cytol.*, 65, 405–413.

[3] Hizume, K., Yoshimura, S. H., and Takeyasu, K. (2004), Atomic force microscopy demonstrates a critical role of DNA superhelicity in nucleosome dynamics, *Cell Biochem. Biophys.*, 40, 249–261.

[4] Sakaue, T. and Lowen, H. (2004), Unwrapping of DNA–protein complexes under external stretching, *Phys. Rev. E Stat. Nonlin. Soft Matter Phys.*, 70, 021801.

[5] Bouchiat, C., Wang, M. D., Allemand, J., Strick, T., Block, S. M., and Croquette, V. (1999), Estimating the persistence length of a worm-like chain molecule from force-extension measurements, *Biophys J.*, 76, 409–413.

[6] Oana, H., Ueda, M., and Washizu, M. (1999), Visualization of a specific sequence on a single large DNA molecule using fluorescence microscopy based on a new DNA-stretching method, *Biochem. Biophys. Res. Commun.*, 265, 140–143.

[7] Washizu, H. and Kikuchi, K. (2006), Electric polarizability of DNA in aqueous salt solution, *J. Phys. Chem. B Condens. Matter Surf. Interf. Biophys.*, 110, 2855–2861.

[8] Herrick, J. and Bensimon, A. (1999), Single molecule analysis of DNA replication, *Biochimie*, 81, 859–871.

[9] Caburet, S., Conti, C., and Bensimon, A. (2002), Combing the genome for genomic instability, *Trends Biotechnol.*, 20, 344–350.

[10] Lee, G. U., Chrisey, L. A., and Colton, R. J. (1994), Direct measurement of the forces between complementary strands of DNA, *Science*, 266, 771–773.

[11] MacKerell, A. D., Jr and Lee, G. U. (1999), Structure, force, and energy of a double-stranded DNA oligonucleotide under tensile loads, *Eur. Biophys J.*, 28, 415–426.

[12] Shin, J. S. and Piercejk, N. A. (2004), Rewritable Memory by Controllable Nanopatterning of DNA, *Nano Lett.*, 4, 905–909.

[13] Yan, H., Zhang, X., Shen, Z., and Seeman, N. C. (2002), A robust DNA mechanical device controlled by hybridization topology, *Nature*, 415, 62–65.

[14] Kidoaki, S. and Yoshikawa, K. (1999), Folding and unfolding of a giant duplex-DNA in a mixed solution with polycations, polyanions and crowding neutral polymers, *Biophys. Chem.*, 76, 133–143.

[15] Katsura, S., Yamaguchi, A., Hirano, K., Matsuzawa, Y., and Mizuno, A. (2000), Manipulation of globular DNA molecules for sizing and separation, *Electrophoresis*, 21, 171–175.

[16] Mihardja, S., Spakowitz, A. J., Zhang, Y., and Bustamante, C. (2006), Effect of force on mononucleosomal dynamics, *Proc. Natl Acad. Sci. USA*, 103, 15871–15876.

[17] Jiang, Y., Qin, F., Li, Y., Fang, X., and Bai, C. (2004), Measuring specific interaction of transcription factor, ZmDREB1A with its DNA responsive element at the molecular level, *Nucleic Acids Res.*, 32, e101.

[18] Yu, J., Sun, S., Jiang, Y., Ma, X., Chen, F., Zhang, G. et al. (2006), Single molecule study of binding force between transcription factor TINY and its DNA responsive element, *Polymer*, 47, 2533–2538.

[19] Yoshida, Y., Nojima, Y., Tanaka, Y., and Kawai, T., (2007), Scanning tunneling spectroscopy of single-strand deoxyribonucleic acid for sequencing, *J. Vac. Sci. Technology* B: Microelectronics and Nanometer Structures, 25, 242–246.

## CHAPTER EIGHT

# SINGLE-MOLECULE PROTEIN MECHANICS

## Contents

## 8.1 PROTEIN-STRETCHING EXPERIMENTS

Protein–stretching experiments together with theoretical simulations are a hot topic in force spectroscopy using AFM. Experiments involve sandwiching a single protein molecule between an AFM probe and a solid substrate and pulling up the probe together with a part of the protein adhered to the probe. If the other end of the protein is somehow immobilized to the substrate, the protein is stretched to two opposing directions, while its 3D structure is gradually unfolded by the tensile force applied by the cantilever. In the process of stretching, the AFM records the tensile force being applied to the protein as the product of cantilever displacement $(d)$

and its force constant $(k)$ and of the distance between the sample surface and the tip of the probe $(D)$. Since $(D - d)$ is equal to the stretched length of the protein, one can construct a force-extension $(F-E)$ curve, which describes the mechanical characteristics of the 3D structure of the protein. If the protein is stretched without much resistance, the $F-E$ curve will be flat until the extension reaches the full length of the polypeptide chain. If the protein contains locally rigid structures, the $F-E$ curve will be characterized by the appearance of one or more force peaks each of them corresponding to the rupture event of a locally rigid structure. Each force peak is followed by a nonlinear increase of the force representing the stretching of the polypeptide chain that is freed from the rigid local domain after its breakdown. The peak value of each force peak gives the tensile strength of the domain being unfolded.

Figure 8.1 gives the schematic view of the protein-stretching event and the notable features of $F-E$ curve corresponding to the elementary steps of protein unfolding.

The protein stretching experiment is otherwise called as 'forced unfolding', or 'mechanical unfolding', and the relationship between the tensile force and the stretching distance is often termed as 'force spectroscopy'.

The basic idea of protein stretching was introduced by Mitsui et al. in 1996 [1]. They immobilized $\alpha$-2-macroglobulin on a gold–coated mica surface through the covalent cross-linker, succinimidylpyridyldithio propionate (SPDP), which linked amino groups on the protein to the gold surface. The opposite surface of the protein was cross-linked to a gold-coated AFM probe, thus covalently sandwiching the protein between the substrate and the probe. Then, the probe–substrate distance was increased to mechanically unfold the protein, and the relationship between the tensile force and the protein extension was obtained. The slope of the $F-E$ curve was interpreted as the stiffness of the protein. The experimental setup and the result of the stiffness data are given in Figure 8.2.

Ikai et al. applied a similar method to stretch carbonic anhydrase II and obtained similar value of stiffness as for $\alpha$-2-macroglobulin [2].

In the above experiment, the protein molecule was pulled from randomly chosen lysine residues on the two opposing sides of the

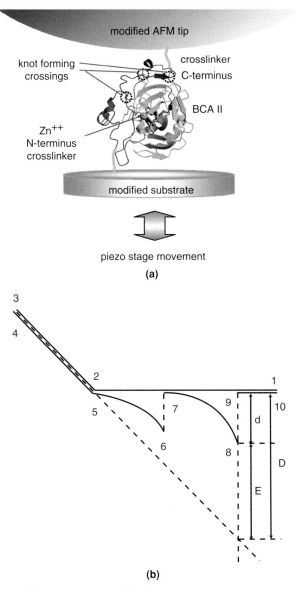

**Figure 8.1** Schematic view of single-protein-stretching experiments on the AFM. (a) Protein molecule, carbonic anhydrase in this case, is sandwiched between the probe and substrate on AFM, and subsequently the distance between them is increased. (b) Typical force curve to be obtained from protein-stretching experiments along the numbered sequence from 1 to 10. Symbols are: $d$ = cantilever displacement, $E$ = protein extension length, and $D$ = piezo motor displacement.

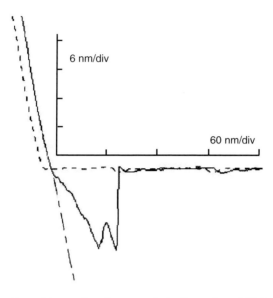

**Figure 8.2** Stretching of $\alpha$-2-macroglobulin using AFM. The large protein with the molecular weight of 720,000 was pulled from two opposing surfaces. Reproduced from the study by Mitsui et al. [1] with permission.

molecules and not necessarily from the two ends of the molecules. The protein was probably extended as a solid spherical material and not as a linear polymer chain. To obtain reproducible $F$–$E$ curves and to probe the local rigidity as a function of stretching length, it is desirable to stretch a protein molecule of known 3D structure from its N- and C-terminals. Alam et al. conducted such an experiment on bovine carbonic anhydrase II [3]. The protein is known to have a knot structure in its C-terminal region (Figure 8.1a), and it had been shown previously by Wang et al. [4] that the protein could only be extended to approximately 20 nm before the covalent cross-link system was severed, with the expected contour length of the protein of 259 amino acid residues being 96 nm.

Alam et al., therefore, constructed a mutant protein that had a replacement of glutamine residue at the 253th position with cysteine residue and used it as a experimental object for knot-free stretching. The mutant protein was stretched up to 70 nm with maximum tensile force of approximately 1 nN before the rupture of covalent cross-linking system. Occasionally, there was a sharp decrease of the force from approximately 1 nN to approximately 200 pN, signifying a structural transition from a tightly folded state (type I) to a more loosely folded state (type II). In fact, they found the coexistence of type I and type II molecules of the same mutant protein in solution after expression in *Escherichia Coli* and purification. The two forms of the mutant protein had almost the same far-UV-CD spectrum and intrinsic fluorescence properties, but a slightly different near-UV-CD spectrum, implying a native-protein-like folding in terms of the secondary structure but a slightly loose tertiary folding. Partial digestion of the C-terminal amino acid residues using carboxypeptidase coupled with MALDI TOF analysis of the molecular weight confirmed that the final sequestration of the C-terminal residues in terms of knot formation was incomplete in type II molecules.

A subsequent SMD simulation of the stretching process of carbonic anhydrase by Ohta et al. [5] strongly suggested that the structural transition from type I to II was accompanied by the concerted breakdown of the centrally located three $\beta$-sheet strands together with forced release of the catalytic $Zn^{2+}$ ion coordinated to three histidine residues on the three strands. Thus, the unusually strong folding force of the protein at the final stage of forced unfolding was considered to have originated from the rupture event of a coordination bond, which is quite strong though not as strong as a covalent bond.

Bacteriorhodopsin is a membrane protein with seven membrane-spanning helices, and much effort has been made to understand and utilize the conversion mechanism of light energy to chemically driven proton pumps. The folding pattern of the seven helices was studied using the forced unfolding method of AFM [6]. Each of the helices was sequentially pulled out of the lipid membrane showing a saw-tooth pattern of force peaks linked with stretching curve of almost randomly coiled chains.

 ## 8.2 INTRAMOLECULAR CORES

Hertadi et al. showed the presence of core rigid structures in OspA protein [7] and holo-calmodulin [8]. The results are given in

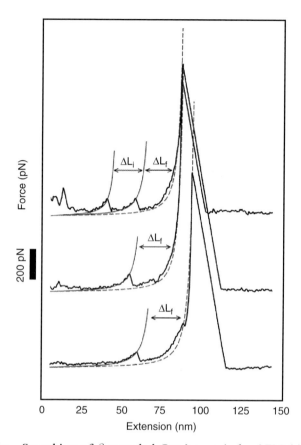

**Figure 8.3** Stretching of β-stranded OspA protein by AFM (a) for the native protein, two force peaks were noted, whereas for two other mutant proteins in (b) and (c), only one peak was observed, reflecting the effect of amino acid substitution (b) and insertion of extra β-sheets (c) in the mid-region of the molecule. Details are provided in the original literature. Reproduced from the study by Hertadi et al. [7] with permission.

Figures 8.3 and 8.4. In both cases of OspA and holo-calmodulin, one or two force peaks were observed in the process of mechanical stretching of the proteins, suggesting that there were rigid core structures inside the molecules that resisted the tensile force up to certain point and then catastrophically fractured into random coils. In the case of apo-calmodulin, which lacked bound

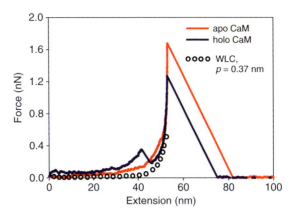

**Figure 8.4** Stretching of apo- and holo-calmodulin for holo-calmodulin, and for apo-calmodulin, and for worm like chain model with persistence length $p = 0.37$ nm. Reproduced from [8] with permission.

$Ca^{2+}$ ions, the extension curve was a smooth one without force peaks, but the force value itself was noticeably higher than that of random-coil stretching. Since apo-calmodulin is made of approximately 60% $\alpha$-helical conformation, mechanical stretching of such conformation is a relatively low-force event, without brittle transition to random coil at least up to 80–90% of the full extension.

Afrin et al. stretched a purely $\alpha$-helical polypeptide based on poly-L-alanine with regular insertion of lysine residues to promote its solubility in water and found that the stretching force curve was very close to the one extracted under denaturing conditions [9]. The result is a strong confirmation that an isolated single $\alpha$-helical chain is easily extended with a low force without brittle transition. The result is seemingly different from the one reported by Idiris et al. [10] on the mechanical stretching of poly-L-glutamate with a high content of $\alpha$-helical conformation in which case the force curve was smooth, but the force was much higher than that observed for either apo-calmodulin or $\alpha$-helical-alanine-based polypeptides. The difference is probably due to the geometrical constraints on the dihedral angles when peptide groups are stretched under tensile conditions without allowance for annealing of the dihedral constraints. Thus, bulky side groups of poly-L-glutamate could have more strongly blocked the relaxation of the

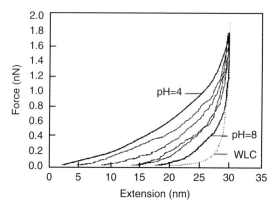

**Figure 8.5**   Stretching of helical peptides: poly-L-glutamate (reproduced from [10] with permission).

dihedral angles from those fit for $\alpha$-helical conformation to those fit for the fully stretched $\beta$-sheet conformation.

 **8.3 STRETCHING OF MODULAR PROTEINS**

Gaub and colleagues started using titin molecule which has a large number of globular units tandemly connected by intervening flexible chains [11]. When the protein was placed on gold-coated mica, a random adhesion took place, and when a gold-coated AFM probe was pushed into the protein layer with a strong force, again a random adhesion of some of the titin chain took place. Then, the probe–substrate distance was increased, and the titin molecules that adhered to the probe on one part and to the substrate on another were stretched out. The molecules that adhered with short interval chains were broken one after the other, and finally the longest chain remained to show the mechanics of single-molecule stretching (Figure 8.6).

There have been numerous examples of stretching mechanics of modular proteins and others. Readers are referred to [12, 13] for a review of this field. The force peaks in the saw-tooth pattern

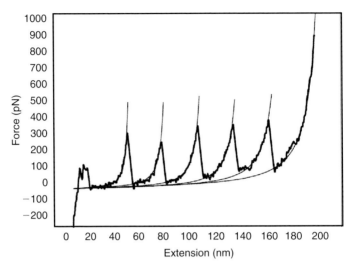

**Figure 8.6** An example of modular-protein stretching. The protein given in this example is the muscle elastic protein, titin (otherwise called connectin). Reproduced from the study by Rief et al. [11] with permission of SCIENCE.

represent the yield force of individual repeating units. In most of the cases, where yield force higher than 50 pN was observed, the molecular basis of the tensile resistance was attributed to the shearing rupture of cooperatively aligned multiple hydrogen bonds between parallel $\beta$-strands. If two $\beta$-sheets are hydrogen bonded in an antiparallel manner, unzipping takes place with a much smaller force, as clearly shown by Brockwell et al. [14] using engineered proteins.

Mechanical stretching of a single protein molecule can reveal the presence of locally rigid structures within the native structure. For example, if the $F-E$ curve of protein stretching has one or more than one force peaks, the protein has a local structure that can only be disrupted after the application of a certain intensity of force. The cause for the appearance of force peaks can be due to any one of the following:

• Intramolecular disulfide bonds or other types of cross-links

• Coordination bonds with a metal(s) ion as the center

- Strongly folded tertiary structure
- Complex formation with ligand molecules
- Strongly hydrogen-bonded core structures

Exposure of the presence of locally rigid structures inside the protein molecules and the identification of such structures in terms of the primary and tertiary structures are one of the most exciting possibilities that nano–biomechanics can present. Development of experimental methods and coordinated computer simulations together with the knowledge from traditional folding studies would bring in a bright prospect in protein science.

 ## 8.4 DYNAMIC STRETCHING

The dynamic method of protein pulling was initiated by Mitsui et al. [15] and later extended by Okajima et al. [16]. Okajima et al. pulled covalently tethered carbonic anhydrase II from the silicon surface by a sinusoidally oscillating AFM probe. They first reproduced a sharp transition from type I to type II form of the protein as reported by Alam et al. [3]. As they recorded the movement of the sinusoidally oscillating cantilever, it was found to be was in phase with the input signal given to vibrate the cantilever. However, as the chain extension length approached the transition region, a gradual phase shift was observed between the input signal and the actual vibration of the cantilever as the output and the cantilever oscillation became highly damped in the transition zone. After the transition from type I to II, the oscillatory behavior of the cantilever was restored to the original one before the transition. Apparently, there was a pulling force acting on the cantilever from the transiently extended protein molecule in the transition zone, which could be ascribed to the force that acts to restore the native 3D conformation of the protein molecule.

Recent development in dynamic stretching can be found in the study by Kawakami et al. and Khatri et al. [17, 18]. They studied several different kinds of macromolecules using an AFM equipped with an oscillatory cantilever, which recorded changes in stiffness of the macromolecules, thereby revealing the frictional-energy dissipation mode within single molecules.

## 8.5 CATCH BOND

When the protein–ligand affinity is affected by the application of a pulling force, it is referred to as the 'catch bond'. The increased affinity is attributed to the conformational change of the protein due to the applied force. It is thus analogous to the change in affinity of an enzyme to its substrate, or of a receptor to its ligand, as the result of binding of an allosteric effector. An example of the catch bond is found in the FimH protein of *E. Coli* fimbrae (pili) [19, 20]. The long and thin fibrous structures extending from the surface of *E. Coli* have adhesive proteins at their ends, which are used to immobilize the bacteria to the nearby surface temporarily. When assayed in a flow chamber, where a gentle flow of water at a constant flow rate could be maintained, bacteria exhibited adhesion and detachment and were, as a consequence, slowly carried down along the water flow. As the flow rate was increased, duration of adhesion time increased and under an even higher flow rate, almost all the bacteria were immobilized on the surface. The effect was considered to have a biological meaning in that bacteria would not be washed away from their natural habitat by a temporarily strong flow. Figure 8.7 shows the result obtained for the catch bond.

## 8.6 PROTEIN-COMPRESSION EXPERIMENTS

The rigidity of a protein molecule can be more directly evaluated by compressing it. Suppose you have two spherical objects made

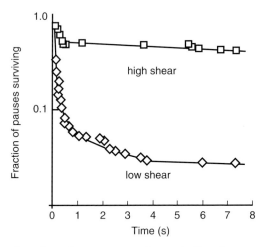

**Figure 8.7** The detatchment rate dramatically decreased under a stronger water flow (upper curve) compared with the result under a slower flow. Reproduced from the study by Thomas et al. [19] with permission.

of either the same or different materials. When the two spheres are compressed against each other with a force normal to the plane of contact, the spheres will be flattened in the contact region, and the degree of flattening depends on the mechanical rigidities (Young's modulus and Poisson's ratio) of the materials with of which the two spheres are made of.

### 8.6.1 Hertz model

The relationship between the normal force and the degree of flattening on the two spheres has been treated by Hertz [21–23]. Figure 8.8 describes the geometrical relationship between the force and the deformation.

Two spheres with radius $R_1$ and $R_2$ are compressed against each other by the applied force $F$, and the force is along the direction of the straight line connecting the centers of the two spheres. The shapes of the two spheres (solid lines) deviated from those of the perfect spheres observed before contact (dotted lines). If the position on the solid lines is expressed as ($x_1$, $y_1$, and $z_1$) for

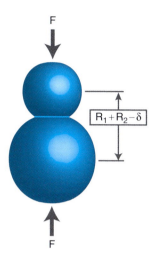

**Figure 8.8**  Hertz model deals with two spheres in contact. Both spheres are deformed, and the definitions of parameters are provided in the text.

sphere 1 and $(x_2, y_2,$ and $z_2)$ for sphere 2 with the directions of three coordinates as given in the figure, deformation for sphere 1 is expressed as $\delta_1 = R_1 - z_1$, and that for sphere 2 is $\delta_2 = R_2 - z_2$. The applied force at the origin, $(0, 0, 0)$, exerts a pressure distribution in the contact region. The functional form for the pressure distribution was assumed to be as indicated below when Hertz derived the relationship between $\delta_1(0, 0, 0)$ and $\delta_2(0, 0, 0)$ using the following function for the pressure distribution over a circular contact area of radius $a_0$ between the two spheres.

$$p(r) = p_0(a_0^2 - r^2)^{1/2}/a_0 \tag{8.1}$$

where $p_0$ is the pressure at the center of the circle.

According to Hertz, the depths of depression $\delta_1(0, 0, 0)$ and $\delta_2(0, 0, 0)$ are related to the applied force through the following equation, where $Y_1$ and $Y_2$, and $\nu_1$ and $\nu_2$ are Young's modulus and Poisson's ratio of materials that make up spheres 1 and 2. In the derivation of the Hertz equation, $\delta_1(0, 0, 0)$ and $\delta_2(0, 0, 0)$ were assumed to be much smaller than $R_1$ and $R_2$ and deformations of the spheres assumed to be only vertical and no lateral extension was allowed (see Appendix 13.9).

$$F = \frac{4\sqrt{R}}{3}\left[\frac{(1-\nu_1^2)}{Y_1} + \frac{(1-\nu_2^2)}{Y_2}\right]^{-1} I^{3/2} = \frac{4\sqrt{R}}{3}Y^* I^{3/2}$$

(8.2)

where

$$\frac{1}{Y^*} = \frac{(1-\nu_1^2)}{Y_1} + \frac{(1-\nu_2^2)}{Y_2} \quad I = \delta = \delta_1 + \delta_2$$

$$\text{and } R = \frac{R_1 R_2}{R_1 + R_2}$$

(8.3)

In the case where a sphere is compressed against a flat surface, one can set either one of $R_1$ or $R_2$ as infinitely large, and in the case where hardness of the two spheres is much different, one can set either one of $Y_1$ or $Y_2$ to infinity. Thus, when a protein molecule (with $R_1$ and $Y_1$) is compressed under an AFM probe, which is much harder than the protein, the above relationship will become

$$F = \frac{4}{3}\frac{Y_1\sqrt{R_1}}{(1-\nu^2)} I^{3/2} = aI^{3/2}$$

(8.4)

where $R_2$ and $Y_2$ are set as infinitely large.

On the basis of this type of reasoning, Radmacher et al. analyzed compression curves of lysozyme adsorbed on mica surface. By assuming the Poisson's ratio to be approximately 0.35, they obtained $Y$ in the range of $500 \pm 200$ MPa [24].

Afrin and Ikai considered this problem in their recent report on the compression of carbonic anhydrase II [25]. In their analysis of the compression data, they applied a recent modification of Hertz model by Tatara, which extended the original model to describe large deformation cases where a spherical sample is allowed to undergo an extensive flattening and lateral extension, both of which were not allowed in the original Hertz model. Tatara developed his model to explain a large deformation of a homogeneous and isotropic rubber sphere by using a constant value of Young's modulus [26–28].

By applying Tatara model, which is explained in the next subsection, to the compression curve of carbonic anhydrase II, Afrin and Ikai found that nearly 50% of the compression curve was fitted by using a constant value of Young's modulus, whereas fitting of Hertz model was possible only in the initial 10% of the compression curve. It is important to note that, in the estimation of Young's modulus from the analysis of compression curves, the elasticity of the sample should be confirmed by checking the extent of overlapping of the approach and the retraction parts of the curves.

Other methods for the measurement of Young's modulus of the protein includes, (1) mechanical stretching of a single actin fiber by Kojima et al. [29], (2) deduction from the bulk compressibility determined from the measurement of the speed of sound wave propagation in protein solution [30], (3) compression of a large number of protein molecules under a surface-force apparatus [31], and (4) vibrational analysis of protein crystals [32]. The results of all these measurements gave values over a wide range of 200 MPa to 10 GPa. The most reliable result was due to Kojima et al., who immobilized an actin filament and measured the relative elongation of the fiber under a given tensile stress. They obtained $Y$ for actin monomer as 2 GPa and that for tropomyosin as 10 GPa. In both cases, strain was less than a few per cent of the original length of the fiber and can be considered the rigidity under small deformation.

A caveat here is when one uses of bulk compressibility, $\kappa$, for the deduction of Young's modulus by using the well-known equation as follows.

$$Y = 3(1 - 2\nu)/\kappa \qquad (8.5)$$

It gives a reasonable value of Young's modulus as long as $\nu$ is not too close to 0.5, but for materials with $\nu$ close to 0.5, the factor $(1-2\nu)$ becomes too close to zero, and a small change in $\nu$ leads to a considerably large change in $Y$. The value of $\nu$ is not known for proteins and usually a value in the range of 0.3–0.35 is assumed, claiming that proteins may have mechanical properties similar to artificial plastics such as polyethylene or polystyrene. However, the above point has not been well established. It is more likely that proteins resemble a rubber ball rather than spherical plastics and that $\nu$ is closer to 0.5 than the previously assumed values. Thus, deduction of $Y$ from Eq. (5) should not be used unless Poisson's ratio of the protein sample is known accurately.

The lower values of $Y$ obtained from compression experiments by using AFM are based on a large-scale deformation of the protein molecules, and high values for actin and tropomyosin come from small-deformation data. The results indicate that the protein is rigid against a small-scale deformation and behaves as softer material for large-scale deformation, the behavior reminiscent of an egg with rigid shell containing soft material of egg white and yolk.

## 8.6.2 Tatara model

Limitations of Hertz model when applied to the real system have been recognized, for example, when two spheres have very different Young's modulus, the contact area cannot be flat and the contact surface is often adhesive. The latter condition was incorporated into the development of JKR model from Hertz model [21].

When indentation is extensive, some of the assumptions in Hertz model must be modified. Tatara developed an indentation model of homogeneous and isotropic sphere sandwiched between two rigid, flat, and parallel plates [26–28]. Compressing force is applied from the upper plate, whereas the lower plate is kept immobile, allowing the sandwiched sphere to be deformed symmetrically from the top and the bottom. Lateral extension of the sphere is considered in the model so that the sphere is allowed to deform such that the distance between the top and bottom plates was less than a half of its original diameter. The apparent depth of indentation measured in AFM instrument is twice the original definition of Hertz indentation, $I_H$, because the sample is deformed both from the top and the bottom. To comply with the definition of the depth of indentation in the original Hertz model, $I_H$ is taken as one half of the measured compression of the sample. Power expansion of the original analytical formula gives the following relationship between $F$ and $I_H$ using $a$ as defined in Eq. (8.4).

$$F = aI_H^{3/2} + \left[\frac{3a^2}{2a_c}\right] I_H^2 + \left[\frac{15a^3}{8a_c^2}\right] I_H^{5/2} \qquad (8.6)$$

The above equation results from the expansion of the following equation, assuming that $Y_2$ for the substrate is much larger than $Y_1$ for the sample.

$$I_H = \left[\frac{F}{a}\right]^{2/3} - \frac{F}{a_c} \quad \text{where} \qquad (8.7)$$

$$\frac{1}{a_c} = \frac{(1+\nu_1)(3-2\nu_1)}{4\pi Y_1 R_1} + \frac{(1+\nu_2)(3-2\nu_2)}{4\pi Y_2 R_2} \qquad (8.8)$$

The coefficients in the above equation can be reduced to contain a single parameter that is similar to the one used in the Hertz model. When Poisson's ratio is assumed to be 0.4, the above equation is reduced to the following one.

$$F = aI_H^{3/2} + 0.337aI_H^2 + 0.0948aI_H^{5/2} \qquad (8.9)$$

As stated above, this result was applied to the analysis of the compression data of bovine carbonic anhydrase II (BCA II) by Afrin et al. [25]. While the application of Hertz model required Young's modulus to continuously change from 70 MPa to 200 MPa as the indentation proceeded deeper into the protein molecule, Tatara model explained the indentation curve up to 50% of the total height of the molecule with a constant Young's modulus of 75 MPa. The Poisson's ratio was kept constant at 0.4 because the change in this ratio would not affect the outcome of Young's modulus to any significant degree.

A forced fitting of Hertz model to the same data range, gave a $Y$ value of approximately 150–200 MPa. Application of the classical Hertz model to small objects such as protein molecules compressed between rigid surfaces should be treated with caution. If the result of indentation done by Radmacher et al. is reexamined in the light of Tatara model, the estimate of 500 MPa was likely to be an overestimate by at least a factor of two to three. Another reason that lysozyme gave a still larger Young's modulus than that of BCA II was that the former has four disulfide bonds in a molecule of less than half of BCA II in molecular weight. It is also possible that the indentation curve of BCA II was fitted by Tatara model only up to 50% of its height. The rigid core of the molecule as probed by the pulling experiment of the same protein may not have been accessible to compression analysis up to 50% of its compression. Unpublished result on serum albumin suggests that the protein is as soft as BCA II in compression analysis.

## 8.7 INTERNAL MECHANICS OF PROTEIN MOLECULES

Experiment of mechanical unfolding of a single protein molecule was expected to lead to a ground breaking view of protein folding, which has been one of the most important and yet most difficult problems in biochemistry. The mechanism of protein folding is studied from both thermodynamical and kinetic point of views. According to Tanford, globular protein molecules are in thermodynamically most stable state, but the difference in Gibbs energy between the native and unfolded is as small as 10–100 kJ/mol, and it has been said that the native protein is 'marginally stable' against unfolded state [33]. The stability of the native protein molecule is the result of several contradicting factors, each contributing to the stabilization or destabilization of the native state. The sum total of all of them when interpreted in terms of Gibbs energy happened to be slightly negative for the native conformation than for the unfolded one under physiological conditions. The contradicting factors include, for example, (1) conformational entropy strongly favoring unfolded state, (2) hydrophobic interactions between non-polar side chains and the following interactions all favoring folded native state, (3) hydrogen bonding between hydrogen donors and acceptors, (4) ion-pair formation, (5) van der Waals interaction between closely packed residues inside the folded structure, and, where applicable (6) disulfide-bond formation also favoring the folded conformation.

The pathway of folding from an open to a closed form is studied using kinetic data and molecular dynamics simulations. There are simpler cases where 'two state transition' theory applies, meaning only the native and the completely unfolded state makes any significant contribution to the thermodynamics of protein folding. Intermediate states, if any, do not accumulate to a measurable degree throughout the folding and unfolding reactions. When the presence of partially folded states is noted in the kinetic study, characterization of any identifiable state will be pursued. In the early stage of kinetic studies of protein folding and unfolding, the presence of well-defined intermediate states was explored, but a currently more popular concept is that of the energy landscape characterized by multiple folding pathways, each polypeptide chain following different pathways to fold into the native state, and thus there may not be accumulation of well-defined intermediate states [34]. Whether mechanical unfolding experiments can contribute to the folding theory is currently a hot topic.

## 8.8 MECHANICAL CONTROL OF PROTEIN ACTIVITY

As briefly mentioned in Chapter 1, it will be interesting if we could control the enzymatic activity of a protein molecule by simply pushing or pulling the molecule to inflict a small deformation in its active site. If the deformation is reversible, the activity can be controlled reversibly and quickly by the application of a tensile or compressive force as long as necessary. Usually enzyme activity is conveniently controlled by the addition of a specific inhibitor or allosteric effector to the enzyme solution, whereas reactivation requires removal of added reagents by, for example, a time-consuming procedure of dialysis.

Kodama et al. reported an attempt to control the intensity of fluorescence emission of green fluorescent protein (GFP) by cyclically repeating the compression and extension of the GFP molecules sandwiched between the substrate and an AFM probe with an attached glass bead [35]. No accurate estimate of the number of GFP molecules under the glass bead was available, but they observed cyclic change of the fluorescence intensity in phase with the mechanical cycle of the cantilever (see Figure 8.9).

For this purpose, they developed a combined system of AFM and confocal fluorescence microscope [36].

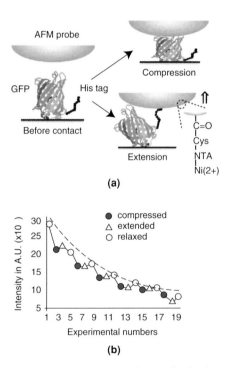

**Figure 8.9**   Fluorescence intensity was diminished when GFP was compressed or extended. Reproduced from the study by Kodama et al. [36] with permission.

## 8.9 COMPUTER SIMULATION OF PROTEIN DEFORMATION

Nanomechanical experiments deal primarily with atoms and molecules that are not visible in the process of experimental manipulation, except in cases where the entire operation was visualized using electron microscopy [37].

In protein-stretching experiments, for example, we like to see the proteins being stretched, exposing their sequential fracture mechanism under tensile force and obtaining a correlation between the force-extension curve and the molecular events observed by our naked eye. We also like to ensure that, in the compression experiment, the protein in focus is right under the probe and is compressed without dodging the probe. Since it is not possible to visualize the molecular events that are observed by our naked eye, computer simulations have become an important tool for the interpretation of the experimental results obtained by manipulating single molecules under various types of probes to bring us to the realm of virtual visualization of atomic and molecular events. Computer simulation based on the atomic coordinate of the sample material has been actively pursued in the analysis of adsorbed atoms and molecules on a solid surface, especially on a silicon surface. In the biological application of nano-technology also, computer simulation is indispensable to interpret the images and force curves obtained by AFM.

One of the early attempts to achieve such objective was made by Schulten and associates using so-called steered molecular dynamics (SMD) simulations on the forced separation process of a biotin–avidin pair [38, 39].

Other examples of SMD simulations of the process in forced unbinding of ligand–receptor pairs or unfolding of protein molecules are listed in the references [40, 41]. Ohta et al. simulated the mechanical unfolding process of carbonic anhydrase II and found that the process should be characterized by the appearance of a few large force peaks and that the molecule becomes totally unfolded only when the last but the largest force peak was overcome [5]. The last peak corresponded to the fracture event of the core structure of the protein made up of three antiparallel strands in $\beta$-sheet configuration, contributing three histidine residues coordinating the active-site zinc ion. The extension length where the final, largest peak appeared in the simulation almost exactly corresponded to the experimentally observed fracture point of the 3D structure of the protein.

# CASE STUDY: CARBONIC ANHYDRASE II

*Rehana Afrin*
Department of Life Science, Tokyo Institute of Technology,
Nagatsuta 4259, Midori-ku, Yokohama, Japan

In this section, we present a summary of nanomechanical work
on bovine carbonic anhydrase II (BCA II). The globular protein,
bovine carbonic anhydrase II (BCA II), has been a target of nanomechanical scrutiny in the Laboratory of Biodynamics at Tokyo
Institute of Technology, Japan.

## Molecular structure

BCA II has 259 amino acid residues and a folded conformation
as shown in Figure 8.10 [42]. It is predominantly a $\beta$-sheet protein, with a minor contribution of $\alpha$-helices. In particular, the
active site is built on the core $\beta$-sheet structure in the central part
of the molecule. Three histidine residues are liganded to hold a
zinc ion ($Zn^{2+}$) in the active center. The terminal end of the

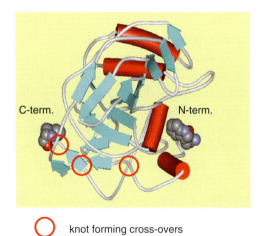

○ knot forming cross-overs

**Figure 8.10**   The crystal structure of bovine carbonic anhydrase II (PDB
code: 1v9e) as determined by Saito et al. [42].

protein is apparently folded to form a pseudo knot as indicated in Figure 8.10. This feature of the protein seemed to present an interesting challenge to test how the AFM-based nanomechanical method responded to the presence of such a structure in a protein molecule. From the number of amino acid residues and assuming the effective length per each amino acid residue to be approximately 0.35–0.37 nm, the total contour length of the molecule was estimated to be 95–100 nm depending on the magnitude of the failure force of the cross-linking system.

## Biological functions

The biological function of BCA II is to catalyze hydration of $CO_2$ and dehydration of $H_2CO_3$, thus facilitating gas–exchange process in the lung, blood stream, and inside the cell. There are more than ten homologs in all the three kingdoms; animals; plants, and bacteria.

## Unfolding studies

BCA II has been regarded as a good example for multistep unfolding reaction, when it is denatured by the addition of denaturant such as guanidinium chloride [43, 44].

## Mechanical stretching from N- and C-termini

To stretch the protein from its two ends, two cysteine residues were inserted at both ends of the molecule, one for each end using the recombinant technology. The surface of crystalline silicon and that of silicon nitride probes for AFM, was chemically functionalized with the silanization reagent, APTES. Both functionalized surfaces were further modified with the covalent cross-linker, SPDP, which has a succinimidyl group that reacts with the amino end of APTES on the substrate, and a pyridyldithio group that forms a covalent bond with cysteine residues on BCA II. When a droplet of the protein solution is placed on a modified substrate and left there for about 30 min, covalent bonds were formed between either end of the protein and the substrate. There was no regulation over the specificity of the end reacting with the substrate, but since the two ends are on opposite sides of the molecule, BCA II molecules were immobilized either with the N-terminal end facing upward or C-terminal facing upward. Thus, an approaching AFM probe was expected to react with the cysteine residue on the dorsal side of the molecule lying on the substrate. Consequently, the experimental system was linked by covalent bonds between the probe and substrate surface, with BCA II molecule sandwiched between the two surfaces.

When the molecule was stretched by a tensile force applied from the AFM cantilever, it extended to approximately 20–30 nm with a small force less than 100 pN and then the force abruptly started to rise with little extension of the molecule [4, 25] as illustrated in Figure 8.11 which was taken as an indication of knot-tightening effect. The force rose to approximately, 2 nN and suddenly the force decreased to zero, indicating a breakdown somewhere in the covalently linked system. The result suggested that the knot became so tightened without allowing for the slippage of the still folded main body of the protein. Only very rarely,

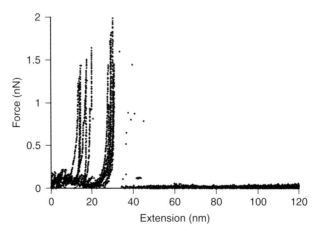

**Figure 8.11** Stretching curve of native BCAII, which had the intact knot structure. The protein was stretched up to approximately 20–30 nm with a force of low intensity, but then the force rapidly increased to higher than 1.5 nN, in which case the covalent system was broken. Reproduced from the study by Afrin et al. [25] with permission.

a continuous stretching of the molecule after breakdown of the knot was observed.

## Knot-free stretching and type I and type II conformers

Alam et al. prepared a mutant protein with glutamine residue at the 253rd position instead of the cysteine residue at the C-terminus [3]. Another cysteine residue was added at the N-terminus. In this case, since the position of 253rd residue is displaced from the knot-forming interaction at the C-terminus, stretching the molecule from this position and from N-terminus was expected to give a knot-free stretching mechanics.

Indeed, this time, stretching of BCA II was much easier than that for the native form of the protein, the extension reaching close to 70–100 nm. Furthermore, by adding the known inhibitor of the enzyme, Alam et al. successfully showed that the mutant protein actually contained two conformational isomers, one called type I having full enzymatic activity and the 3D conformation almost exactly the same as the native protein (PDB 1v9i) [42], and the other conformer called type II, which was enzymatically inactive and not crystallizable. Type II had a highly folded conformation as long as the optical properties such as CD or fluorescence spectra were concerned, suggesting that its secondary structure folding is almost complete, but lacked some special packing of the secondary structures into the native 3D structure.

A partial digestion of the C-terminal residues by a carboxypeptidase treatment followed by MALDI-TOF analysis of the molecular weight of the protein indicated that the C-terminal region of type II was not folded as tightly as in the native protein. It was suggested that, in type II, the knot structure was not formed, leaving the last compacting step of folding incomplete.

In Figure 8.12, the force curves of type I and II conformers are shown as upper and lower curves. They have quite different stiffness as indicated by the slope of the $F–E$ curves. The curves for type I corresponded the knot-free stretching, but type I completely folded conformation, with the formation of the knot as revealed by x-ray crystallography, and it showed a higher stiffness than type II during the unfolding process.

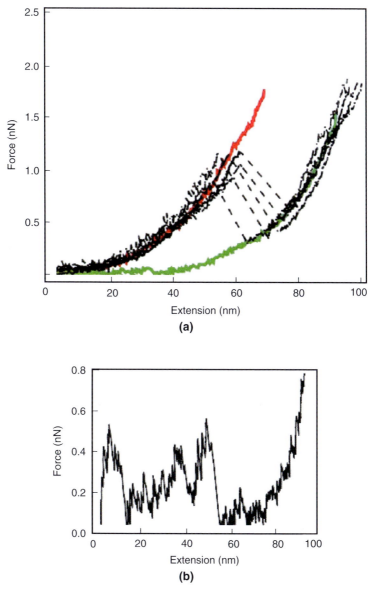

**Figure 8.12** Knot-free stretching of mutant BCA II (Gln253Cys)
a) experimental curves of type I (upper) and type II (lower) stretching.
Transition from type I to type II was occasionally observed. b) Stretching curve obtained as the result of SMD simulation of type I protein.
Reproduced from Refs. [3, 5] with permissions.

### Inhibitor binding

Inhibitor binding to the native as well as to type I conformers made them softer in the initial 30–40 nm of stretching, but then the force reached a level that resulted in covalent bond breaking. It was most prominently shown for type I [3] as given in Figure 8.13. The result was in agreement with the previous observation of the change in the thermal factor in x-ray crystallography before and after inhibitor binding. After inhibitor binding, the thermal factor increased in the peripheral region of the molecule, but decreased in its central region.

### Stretching of partially denatured forms

Afrin et al. studied the stretching curves of partially denatured forms of BCA II [25]. Figure 8.14 shows the result of stretching of partially denatured protein molecules in the presence of 2M GdmCl. The protein was most often extensible to its full length, indicating that the knot structure had been destroyed.

In the above examples of protein stretching, the existence of locally rigid structures has been identified, but the interpretation of the result obtained from chain stretching in terms of mechanical rigidity using numerical expression of Young's modulus is difficult and highly model dependent at best [45], though not impossible.

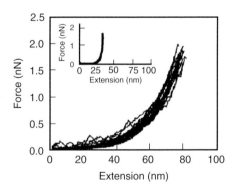

**Figure 8.13**   Stretching curve of type I before (main panel) and after (inset panel) the reaction with an inhibitor. Reproduced from the study by Alam et al. [3] with permission.

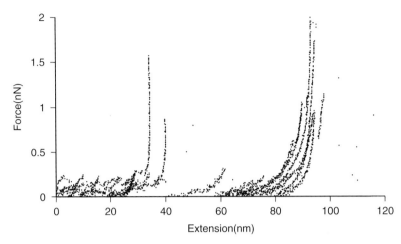

**Figure 8.14**  Stretching of partially denatured forms of BCAII. Reproduced from the study by Afrin et al. [25] with permission.

A better way to characterize the rigidity of a protein molecule numerically is to compress it and analyze the force curve according to the established method in macroscopic mechanics.

## Compression experiment for obtaining Young's modulus

The protein was compressed on the silicon surface by using an AFM probe. The native form of the protein with cysteine residues at N- and C-termini was immobilized on the APTES- and SPDP-functionalized silicon surface and was compressed with a similarly modified silicon nitride AFM probe [25]. The compression curve was given in Figure 8.15 in terms of the Hertzian approach distance, which is one half of the apparent depth of compression and the applied force.

The compression curve was analyzed according to the Hertz model and Tatara model as described previously. They found a good fitting of Tatara model up to 50% of deformation using a constant value of Young's modulus of 75 MPa. Young's modulus of partially denatured and completely denatured protein was also measured.

Figure 8.16 gives the result of computer simulation of compression process of BCAII at 300 K [46]. The calculation was

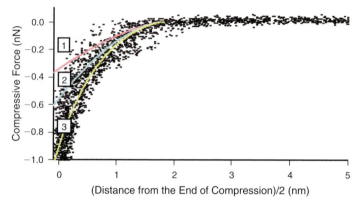

**Figure 8.15** Experimental data and fitting curves in the compression experiment of BCA II. Filled dots: experimental data; Curve 1: Fitting curve of Hertz model; Curve 2: fitting curve of Tatara model; Curve 3: exponetial fitting curve to the experimental data. Reproduced from the study by Afrin et al. [25] with permission.

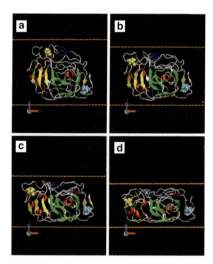

**Figure 8.16** Result of computer simulation of compression process of BCAII in vacuum at 300 K. Four snapshots from the compression process are given from a to d. Reproduced from the study by Tagami et al. [46] with permission.

performed in the absence of water. The compression curve shows detailed force profiles, which are yet to be analyzed and verified experimentally.

Compression experiment in 2M GdmHCL showed that, with the height of 3.5–5 nm, the protein was not expanded as much as the completely denatured form, which had the height of approximately 9 nm.

## Bibliography

[1] Mitsui, K., Hara, M., and Ikai, A. (1996), Mechanical unfolding of alpha2-macroglobulin molecules with atomic force microscope, *FEBS Lett.*, 385, 29–33.

[2] Ikai, A., Mitsui, K., Furutani, Y., Hara, M., McMurty, J., and Wong, K. P. (1997), Protein stretching II1: results for carbonic anhydrase, *Jpn J. Appl. Phys.*, 36, 3887–3893.

[3] Alam, M. T., Yamada, T., Carlsson, U., and Ikai, A. (2002), The importance of being knotted: effects of the C-terminal knot structure on enzymatic and mechanical properties of bovine carbonic anhydrase II, *FEBS Lett.*, 519, 35–40.

[4] Wang, T., Arakawa, H., and Ikai, A. (2001), Force measurement and inhibitor binding assay of monomer and engineered dimer of bovine carbonic anhydrase B, *Biochem. Biophys. Res. Commun.*, 285, 9–14.

[5] Ohta, S., Alam, M. T., Arakawa, H., and Ikai, A. (2004), Origin of mechanical strength of bovine carbonic anhydrase studied by molecular dynamics simulation, *Biophys. J.*, 87, 4007–4020.

[6] Janovjak, H., Muller, D. J., and Humphris, A. D. (2005), Molecular force modulation spectroscopy revealing the dynamic response of single bacteriorhodopsins, *Biophys. J.*, 88, 1423–1431.

[7] Hertadi, R., Gruswitz, F., Silver, L., Koide, A., Koide, S., Arakawa, H. et al. (2003), Unfolding mechanics of multiple OspA substructures investigated with single molecule force spectroscopy, *J. Mol. Biol.*, 333, 993–1002.

[8] Hertadi, R. and Ikai, A. (2002), Unfolding mechanics of holo- and apocalmodulin studied by the atomic force microscope, *Protein Sci.*, 11, 1532–1538.

[9] Afrin, R., Takahashi, I., Ohta, S., and Ikai, A. (2003), Mechanical unfolding of alanine based helical polypeptide: experiment versus simulation, Presented at the 47th Annual Meeting of the American Biophysical Society, San Antonio, Texas, USA. March 1–5, 2003.

[10] Idiris, A., Alam, M. T., and Ikai, A. (2000), Spring mechanics of alpha-helical polypeptide, *Protein Eng.*, 13, 763–770.

[11] Rief, M., Gautel, M., Oesterhelt, F., Fernandez, J. M., and Gaub, H. E. (1997), Reversible unfolding of individual titin immunoglobulin domains by AFM, *Science*, 276, 1109–1112.

[12] Carrion-Vazquez, M., Oberhauser, A. F., Fisher, T. E., Marszalek, P. E., Li, H., and Fernandez, J. M. (2000), Mechanical design of proteins studied by single-molecule force spectroscopy and protein engineering, *Prog. Biophys. Mol. Biol.*, 74, 63–91.

[13] Müller, D. J., Heymann, J. B., Oesterhelt, F., Móller, C., Gaub, H., Buldt, G. et al. (2000), Atomic force microscopy of native purple membrane, *Biochim. Biophys. Acta.*, 1460, 27–38.

[14] Brockwell, D. J., Paci, E., Zinober, R. C., Beddard, G. S., Olmsted, P. D., Smith, D. A. et al. (2003), Pulling geometry defines the mechanical resistance of a beta-sheet protein, *Nat. Struct. Biol.*, 10, 731–737.

[15] Mitsui, K., Nakajima, K., Arakawa, H., Hara, M., and Ikai, A. (2000), Dynamic measurement of single protein's mechanical properties, *Biochem. Biophys. Res. Commun.*, 272, 55–63.

[16] Okajima, T., Arakawa, H., Alam, M. T., Sekiguchi, H., and Ikai, A. (2004), Dynamics of a partially stretched protein molecule studied using an atomic force microscope, *Biophys. Chem.*, 107, 51–61.

[17] Kawakami, M., Byrne, K., Brockwell, D. J., Radford, S. E., and Smith, D. A. (2006), Viscoelastic study of the mechanical unfolding of a protein by AFM, *Biophys. J.*, 91, L16–L18.

[18] Khatri, B. S., Kawakami, M., Byrne, K., Smith, D. A., and McLeish, T. C. (2007), Entropy and barrier-controlled fluctuations determine conformational viscoelasticity of single biomolecules, *Biophys. J.*, 92, 1825–1835.

[19] Thomas, W., Forero, M., Yakovenko, O., Nilsson, L., Vicini, P., Sokurenko, E. et al. (2006), Catch-bond model derived from

allostery explains force-activated bacterial adhesion, *Biophys J.*, 90, 753–764.

[20] Thomas, W. (2006), For catch bonds, it all hinges on the interdomain region, *J. Cell Biol.*, 174, 911–913.

[21] Johnson, K. L. (1985), Chapter 4 in 'Contact Mechanics', Cambridge University Press, Cambridge UK.

[22] Hertz, H. (1882), Über die Beruhrung fester elastischer Korper, *J. Reine und Angewandte Mathematik*, 92, 156–171.

[23] Landau, L. D. and Lifshitz, E. M. (1986), Theory of Elasticity Butterworth Heinemann, Oxford UK.

[24] Radmacher, M., Fritz, M., Clevel, J. P., Walters, D. A., and Hansma, P. K. (1994), Imaging adhesion forces and elasticity of lysozyme adsorbed on mica with the atomic force microscope, *Langmuir*, 10, 3809–3814.

[25] Afrin, R., Alam, M. T., and Ikai, A. (2005), Pretransition and progressive softening of bovine carbonic anhydrase II as probed by single molecule atomic force microscopy, *Protein Sci.*, 14, 1447–1457.

[26] Tatara, Y. (1989), Extensive theory of force-approach relations of elastic spheres in compression and impact, *J. Eng. Mater. Tech.*, 111, 163–168.

[27] Tatara, Y. (1991), On compression of rubber elastic sphere over a large range of displacements – Part 1: theoretical study, *J. Eng. Mater. Tech.*, 113, 285–291.

[28] Tatara, Y., Shima, S., and Lucero, J. C. (1991), On compression of rubber elastic sphere over a large range of displacements – Part 2: comparison of theory and experiment, *J. Eng. Mater. Tech.*, 113, 292–295.

[29] Kojima, H., Ishijima, A., and Yanagida, T. (1994), Direct measurement of stiffness of single actin filaments with and without tropomyosin by in vitro nanomanipulation, *Proc. Natl Acad. Sci. USA*, 91, 12962–12966.

[30] Tachibana, M., Koizumi, H., and Kojima, K. (2004), Temperature dependence of microhardness of tetragonal hen-egg-white lysozyme single crystals, *Phys. Rev. E Stat. Nonlin. Soft Matter Phys.*, 69, 051921–051924.

[31] Suda, H., Sugimoto, M., Chiba, M., and Uemura, C. (1995), Direct measurement for elasticity of myosin head, *Biochem. Biophys. Res. Commun.*, 211, 219–225.

[32] Morozov, V. N. and Morozova, T. Ya. (1981), Viscoelastic properties of protein crystals: triclinic crystals of hen egg white lysozyme in different conditions, *Biopolymers*, 20, 451–467.

[33] Tanford, C. (1968), Protein denaturation, (1968), *Adv. Protein Chem.*, 23, 121–282; *ibid.* (1970), 24, 1–95.

[34] Baldwin, R. L. and Rose, G. D. (1999), Folding intermediates and transition states, *Trends Biochem. Sci.*, 24, 26–33, 77–83.

[35] Kodama, T., Ohtani, H., Arakawa, H., and Ikai, A. (2005), Mechanical perturbation-induced fluorescence change of green fluorescent protein, *Appl. Phys. Lett.*, 86, 043901-1–043901-3.

[36] Kodama, T., Ohtani, H., Arakawa, H., and Ikai, A. (2004), Development of confocal laser scanning microscope/atomic force microscope sytem for force curve measurement, *Jpn J. Appl. Phys.*, 43, 4580–4583.

[37] Kondo, Y. and Takayanagi, K. (2000), Synthesis and characterization of helical multi-shell gold nanowires, *Science*, 289, 606–608.

[38] Izrailev, S., Stepaniants, S., Balsera, M., Oono, Y., and Schulten, K. (1997), Molecular dynamics study of unbinding of the avidin-biotin complex, *Biophys. J.*, 72, 1568–1581.

[39] Isralewitz, B., Izrailev, S., and Schulten, K. (1997), Binding pathway of retinal to bacterio-opsin: a prediction by molecular dynamics simulations, *Biophys. J.*, 73, 2972–2979.

[40] Gao, M., Craig, D., Vogel, V., and Schulten, K. (2002), Identifying unfolding intermediates of FN-III(10) by steered molecular dynamics, *J. Mol. Biol.*, 323, 939–950.

[41] Gao, M., Lu, H., and Schulten, K. (2001), Simulated refolding of stretched titin immunoglobulin domains, *Biophys. J.*, 81, 2268–2277.

[42] Saito, R., Sato, T., Ikai, A., and Tanaka, N. (2004), Structure of bovine carbonic anhydrase II at 1.95 A resolution, *Acta Crystallogr. D: Biol. Crystallogr.*, 60, 792–795.

[43] Wong, K. P. and Tanford, C. (1973), Denaturation of bovine carbonic anhydrase B by guanidine hydrochloride. A process involving separable sequential conformational transitions, *J. Biol. Chem.*, 248, 8518–8523.

[44] Lindgren, M., Svensson, M., Freskgard, P. O., Carlsson, U., Jonasson, P., Martensson, L. G. et al. (1995), Characterization of a folding intermediate of human carbonic anhydrase II: probing

local mobility by electron paramagnetic resonance, *Biophys. J.*, 69, 202–213.

[45] Ikai, A. (2005), Local rigidity of a protein molecule, *Biophys. Chem.*, 116, 187–191.

[46] Tagami, K., Tsukada, M., Afrin, R., Sekiguchi, H., and Atsushi Ikai, A. (2006), Discontinuous force compression curve of single bovine carbonic anhydrase molecule originated from atomistic slip, *e-J. Surf. Sci. Nanotech.*, 4, 552–558.

## CHAPTER NINE

# MOTION IN NANO-BIOLOGY

## Contents

## 9.1 CELL MOVEMENT AND STRUCTURAL PROTEINS

Eukaryotic cells can change their shape and move due to the 3D network structure called the cytoskeleton. The cytoskeleton is composed of three layers of protein filaments; actin filaments (micro- or thin filaments), microtubules, and intermediate filaments (IF). Their properties are summarized in Table 9.1.

The rigidity of these filamentous structures is of basic importance for understanding the capacity of the cell to maintain its shape and movement. There are several expressions for the rigidity of filamentous structures, all of which are related to the modulus of elasticity (Young's modulus, $Y$).

**Longitudinal stiffness, $k$:** The force required to produce a unit elongation in the longitudinal direction of the sample. For a beam of the cross-sectional area $A$ and length $L$, $k = YA/L$. The inverse of stiffness is defined as flexibility. Knowing $A$ and $L$ of the rod, one obtains the value of $Y$.

**Table 9.1**  Properties of cytoskeletal proteins and filaments [1].

| Proteins | Subunit MW | No. of Protofilaments | Diameter (nm) | Cross-sectional area (nm$^2$) |
|----------|-----------|----------------------|---------------|------------------------------|
| Actin | 45,000 | 2 | 5 | 19 |
| Tubulin | 50,000 | 1–3 | 25 | 200 |
| Intermediate filament | 40,000– 180,000 | 8 | 10 | 60 |
| Coiled coil | — | 2 | 2 | 1.9 |

**Flexural rigidity:**   The moment of a force (force couple) required to bend a rigid beam to a unit curvature, $\kappa$, is equal to $1/R$, where $R$ is the radius of curvature. According to the basic equation of beam-bending mechanics

$$\kappa = \frac{1}{R} = \frac{M}{YI} \tag{9.1}$$

$M$ for the unit curvature is equal to $YI$ and called 'flexural rigidity'. It is a measure of the resistance of a beam to bending; the larger the flexural rigidity, the smaller is the curvature, $\kappa = 1/R$. By dividing $\kappa$ by $I$, one can obtain the value of Young's modulus, $Y$.

**Torsional rigidity:**   The angle of twist, $\phi$, of a linearly elastic beam of length $L$ is related to the torque $T$ and is inversely proportional to 'torsional rigidity', $\tau$, which is equal to $GI_P$, where $G$, and $I_P$ are the rigidity modulus and the polar cross-sectional moment of inertia (defined as $\int r^2(2\pi r dr)$ where $r$ is the radial distance from the center of the cross-sectional area) of the beam, respectively.

$$\phi = \frac{TL}{\tau} = \frac{TL}{GI_P}, \quad I_P = \int A dA = \int r^2(2\pi r dr) \tag{9.2}$$

$I_P$ is $2I$ for axi-symmetric cross-sections. For example, $I$ for a cylindrical rod of radius $r$ is $r^4/4$, whereas $I_P = r^4/2$. Estimation of Young's modulus from the torsional rigidity can be done by calculating $I_P$ for a given cross-section of the sample rod and obtaining the rigidity modulus $G = Y/[2(1 + \nu)]$. For a macroscopic sample, $T$ and $\phi$ are directly measurable, whereas for molecular level samples, measurement of the rotational fluctuation $< \theta^2 >$ under fluorescence microscope is used [2]. The energy of rotational fluctuation is related to the thermal energy through the equipartition theorem *i.e.* $\tau < \theta^2 > /2L = k_B T/2$, where $\tau$ is the torsional rigidity per unit length of the filament and $L$ is the length of the filament.

Table 9.2 gives the estimated values of $Y$ for actin-based filamentous structures from the measurements of three types of mechanical constants.

Living cells are motile, moving from one place to another, either swimming in liquid or crawling on a solid surface. Most of the animals have muscular tissues that are developed for locomotion on the land, in the water, as well as in air. Some of them even crawl underground. Flagellated cells such as *Escherichia coli* or sperms can swim in aqueous media, searching for food and avoiding hazardous environments to them. Some species of mycoplasma glide over the glass surface very elegantly using a fair number of leg-like protrusions from a restricted part of their body. To understand the movement of these organisms in liquid medium, we need to understand a certain level of single molecule mechanics.

**Table 9.2** Young's moduli (GPa) of actin and thin-filament proteins [1].

| Type of measurement | Thin filament (GPa) | Actin–tropomyosin (GPa) | Actin alone (GPa) |
| --- | --- | --- | --- |
| Longitudinal stiffness | 2.3 | 2.8 | 2.3 |
| Flexural rigidity | — | 2.0 | 1.3, 2.6 |
| Torsional rigidity | — | — | 1.5 |

**9.2 MUSCLE AND MOTOR PROTEINS**

Muscle contraction is the result of sliding action between two filamentous protein structures, one made of myosin and the other actin and tropomyosin. The lengths of the two filaments are precisely controlled. In myosin filaments, approximately 200 myosin molecules of $MW = 480,000$ are assembled into a fiber, which is approximately $1.6\,\mu m$ long for rabbit [3], and in actin filaments, G-actin molecules of $MW = 42,000$ are assembled into double-helical filaments of dimeric F-actins of approximately $1.16\,\mu m$ long (thin-filament lengths were significantly different among bovine, rabbit, and chicken, with lengths of $1.28\,\mu m-1.32\,\mu m$, $1.16\,\mu m$, and $1.05\,\mu m$, respectively [4]). The basic assembly of myosin and actin filaments is given in Figure 9.1. The much studied two head groups of myosin molecules continuously bind and unbind to and from the actin fiber. During the unbinding period, the myosin head makes a Brownian motion, moving back and forth over an actin fiber, but, since the forward-going propensity is higher, the myosin head eventually forms a bond with an actin molecule lying ahead at a distance of approximately 8 nm, securing a crucial advancement of muscle-contraction steps. When one molecule of ATP is hydrolyzed during this process, a myosin molecule makes an 8-nm advancement by breaking off from the old binding site and finding a new one. Only a small force of

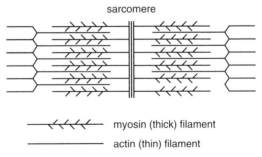

**Figure 9.1** A schematic image of actin containing thin filaments and myosin containing thick filaments in muscle. The cross-bridge structure between the two kinds of filaments discovered by H. E. Huxley provides the force for muscle contraction.

2–5 pN was needed to forcefully unbind the bond between actin and myosin when optical tweezers were used [5]. It is reasonable that the unbinding force is much smaller than the force needed to dissociate antigen–antibody complexes because the former is expected to repeat binding and unbinding so many times at room temperature, whereas the latter is designed to stay associated for a prolonged duration of time.

Discovery of the cross-bridge structure between the actin-containing thin filament and the myosin-containing thick filament laid foundation for the development of mechano-chemical theory of muscle contraction. The rotating cross-bridge hypothesis is based on the following three findings and hypothesis [1].

• Lymn–Taylor scheme of attachment and detachment of myosin from the thin actin filament as regulated by ATP hydrolysis.

• The hypothesis that a myosin molecule has the swinging arm-lever structure that can amplify a small, angström-level conformational change due to ATP hydrolysis to a conformational change due to greater movement of myosin head.

• The power stroke model which assumes the presence of elastic element within the cross-bridge that can store strain energy during the cyclic power stroke.

It is now believed that two other kinds of motor proteins, *i.e.*, kinesin and dynein, function on basically the same mechanism. The simplified notion of the rotating cross-bridge model is illustrated in Figure 9.2.

Kinesin is found to stick to the 'rail', *i.e.*, microtubules, during its processive movement without dissociation. A single kinesin, or a small number of kinesin molecules can carry the load for tens of micrometers along the microtubule. In contrast, the association of myosin with the actin filament is switched on and off; thus a large number of myosin molecules must be bundled together to perform continuous sliding over thin filaments. Myosin is thus termed as nonprocessive. The level of force generated by one stroke of myosin molecule is over the range of 6–9 pN. Some of the large muscle fibers have approximately $10^9$ myosin molecules;

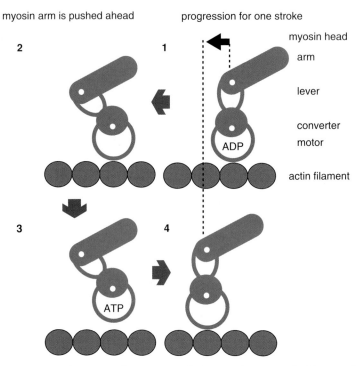

myosin arm is pushed ahead       progression for one stroke

2                                 1

**Figure 9.2** A schematic illustration of the rotating cross-bridge model. Reproduced from [1] with permission.

the collective force generated by a single myosin fiber can be up to mN range.

The speed of motor-protein movement ranges from 100 nm/s to close to 60,000 nm/s in the case of myosin-based systems and from 20 nm/s to 2000 nm/s for kinesin-based systems. There are some systems where motor proteins move in opposite directions from other systems.

## 9.3 SINGLE-MOTOR MEASUREMENTS

Several single-molecule techniques have been used to measure the force required for a single step of motor-protein movement and the distance of the single step.

**Glass-rod cantilever:** A glass capillary can be pulled into a thin rod less than 1 $\mu$m diameter and a few cm long. The spring constant of such a long and thin glass rod can be in the range of a few pN/nm. Such a thin glass rod was used to pull a single actin fiber under a microscope, and the tensile force and the corresponding elongation of the fiber was measured [6]. For a few per cent of elongation against the original length of the fiber, the Young's modulus of an actin molecule was calculated as approximately 2.5 GPa.

**Optical tweezers:** According to Howard [1], the maximum force required for kinesin to perform work was measured by using laser tweezers to be 6 pN with a working distance of 8 nm, *i.e.*, the work that can be done is 48 pN·nm, which is approximately 50% of the maximum work of 100 pN·nm that can be generated by the hydrolysis of a single molecule of ATP ($0.1 \times 10^{-18}$ J $\times$ $6 \times 10^{23} = 60$ kJ). By using optical tweezers system, the following values have been obtained. First, a single molecule of ATP synthetase generates torques as high as 20–40 pN·nm [7], whereas an RNA polymerase can generate forces up to 25 pN [8] and a DNA polymerase 34 pN [9].

## 9.4 FLAGELLA FOR BACTERIAL LOCOMOTION

The flexural rigidity of bacterial flagella was $2.2–4 \times 10^{-24}$ N·m$^2$, which gave a value of 0.5–0.9 GPa for the Young's modulus, assuming the inner $r_i$ and outer $r_o$ diameters of 8 and 10 nm, respectively, where $I_P$ for a cylindrical tube is given as $I_P = \pi(r_o^4 - r_i^4)/2$ [10]. Somewhat larger values of $G = 0.5$ GPa and $Y = 1.5$ GPa were reported when hydrodynamic method was used [11].

## 9.5 MYCOPLASMA GLIDING

Mycoplasma, besides viruses and phages, is a group of microorganisms that has the smallest size. One group of them called

*Mycoplasma mobile* can glide over a glass surface in a highly graceful manner at a speed of a few micrometers per second. The pear-shaped body has a large number of leg-like protrusions as shown in Figure 9.3, which are supposed to provide the driving force for the gliding movement [12]. Miyata et al. measured the force required to prevent the movement of the organism from the focal point of the laser trap. To do so, they immobilized a latex bead on the back of a mycoplasma and kept it in the center of the laser trap. As the organism tried to move away from the trap, they measured the critical force of 26–28 pN as the maximum force generated by the organism on a glass surface. Miyata and Petersen further identified the leg protein [13]. Although the precise mechanism

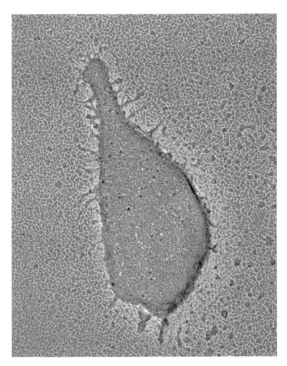

**Figure 9.3**  The scanning electron micrograph of *mycoplasma mobile*. The pear-shaped body as surrounded by spikes, which are considered to be 'legs' to glide over the glass surface. Reproduced from the study by Miyata and Petersen [13] with permission.

of locomotion in this case is not known yet, it seems to be clearly different from other types of motor proteins [14].

## 9.6 MECHANICS AND EFFICIENCY OF MOTOR PROTEINS

The efficiency of motor proteins has been a focus of biophysical studies in an effort to compare their efficiency with that of man-made engines. Kinosita and his colleagues [15, 16, 7] have experimentally shown that the rotary engine of ATP synthetase works at nearly 100% energy efficiency. The proposal and experimental verification of the rotary mode of ATPase action was hailed as a great achievement [17, 18]. The efficiency of the enzyme was estimated against the free energy of hydrolysis of ATP in comparison with the number of protons transported across the lipid membrane.

## Bibliography

[1] Howard, J. (2001), 'Mechanics of Motor Proteins and the Cytoskeleton', Sinaur Associates, Sunderland, MA.

[2] Tsuda, Y., Yasutake, H., Ishijima, A., and Yanagida, T. (1996), Torsional rigidity of single actin filaments and actin-actin bond breaking force under torsion measured directly by in vitro micromanipulation, *Proc. Natl. Acad. Sci. USA*, 93, 12937–12942.

[3] Podlubnaia, Z. A., Latsabidze, I. L., and Lednev, V. V. (1989), The structure of thick filaments on longitudinal sections of rabbit psoas muscle, *Biofizika*, 34, 91–96.

[4] Ringkob, T. P., Swartz, D. R., and Greaser, M. L. (2004), Light microscopy and image analysis of thin filament lengths utilizing dual probes on beef, chicken, and rabbit myofibrils, *J. Anim. Sci.*, 82, 1445–1453.

[5] Ishijima, A., Kojima, H., Funatsu, T., Tokunaga, M., Higuchi, H., Tanaka, H., et al. (1998), Simultaneous observation of individual ATPase and mechanical events

by a single myosin molecule during interaction with actin, *Cell*, 92, 161–171.

[6]  Kojima, H., Ishijima, A., and Yanagida, T. (1994), Direct measurement of stiffness of single actin filaments with and without tropomyosin by in vitro nanomanipulation, *Proc. Natl. Acad. Sci. USA*, 91, 12962–12966.

[7]  Yasuda, R., Noji, H., Kinosita, K., Jr, and Yoshida, M. (1998), F1-ATPase is a highly efficient molecular motor that rotates with discrete 120 degree steps, *Cell*, 93, 1117–1124.

[8]  Wang, M. D., Schnitzer, M. J., Yin, H., Landick, R., Gelles, J., and Block, S. M. (1998), Force and velocity measured for single molecules of RNA polymerase, *Science*, 282, 902–907.

[9]  Wuite, G. J., Smith, S. B., Young, M., Keller, D., and Bustamante, C. (2000), Single-molecule studies of the effect of template tension on T7 DNA polymerase activity, *Nature*, 404, 103–106.

[10]  Fujime, S., Maruyama, M., and Asakura, S. (1972), Flexural rigidity of bacterial flagella studied by quasielastic scattering of laser light, *J. Mol. Biol.*, 68, 347–359.

[11]  Hoshikawa, H. and Kamiya, R. (1985), Elastic properties of bacterial flagellar filaments. II. Determination of the modulus of rigidity, *Biophys. Chem.*, 22, 159–166.

[12]  Miyata, M., Ryu, W. S., and Berg, H. C. (2002), Force and velocity of mycoplasma mobile gliding, *J. Bacteriol.*, 184, 1827–1831.

[13]  Miyata, M. and Petersen, J. D. (2004), Spike structure at the interface between gliding Mycoplasma mobile cells and glass surfaces visualized by rapid-freeze-and-fracture electron microscopy, *J. Bacteriol.*, 186, 4382–4386.

[14]  Uenoyama, A. and Miyata, M. (2005), Gliding ghosts of Mycoplasma mobile, *Proc. Natl. Acad. Sci. USA*, 102, 12754–12758.

[15]  Kinosita, K., Jr (1999), Real time imaging of rotating molecular machines, *FASEB J.*, Suppl. 2, S201–208.

[16]  Kinosita, K., Jr, Adachi, K., and Itoh, H. (2004), Rotation of F1-ATPase: how an ATP-driven molecular machine may work, *Annu. Rev. Biophys. Biomol. Struct.*, 33, 245–268.

[17] Boyer, P. D. (1997), The ATP synthase – a splendid molecular machine, *Annu. Rev. Biochem.*, 66, 717–749.

[18] Noji, H., Yasuda, R., Yoshida, M., and Kinosita, K., Jr (1997), Direct observation of the rotation of F1-ATPase, *Nature*, 386, 299–302.

## CHAPTER TEN

# CELL MECHANICS

## Contents

## 10.1 CHANGES IN SHAPE OF RED BLOOD CELL

Human red blood cells (RBCs) have a well-known discoidal biconcave shape under physiological conditions. The shape of the RBC called for a great deal of interest and attention from both experimental and theoretical investigators. Basically, it is a bag filled with a slightly more viscous liquid than water having a volume of $94 \, \mu m^3$ and a surface area of $135 \, \mu m^2$ in an isotonic 300 milliosmol saline. For a sphere of surface area of $135 \, \mu m^2$, radius would be $3.28 \, \mu m$, and the corresponding volume would be $148 \, \mu m^3$ [1]. Therefore, the unswollen native RBC is approximately 64% full. With this

2/3 filled condition the cell has the well-known biconcave shape and diameter of 8 $\mu$m. The change in the shape of erythrocyte was discussed in Refs. [2, 3].

With regard to the average shape of a RBC, Evans and Fung carefully measured the size and determined the shape of cells and gave the following function to reproduce the cell shape in three dimension [1].

$$Z = \pm R_0 \left[ 1 - \frac{X^2 + Y^2}{R_0^2} \right]^{1/2}$$
$$\left[ C_0 + C_1 \frac{X^2 + Y^2}{R_0^2} + C_2 \left( \frac{X^2 + Y^2}{R_0^2} \right)^2 \right] \quad (10.1)$$

Values of parameters are: $R_0 = 3.91\,\mu$m, $C_0 = 0.207161$, $C_1 = 2.002558$, and $C_2 = -1.122762$, where $2R_0$ is the average cell diameter in the axial direction. Investigations were performed on varying initial cell-diameter values ranging from 7 to 8.5 $\mu$m. Figure 10.1 shows the original equilibrium shape of the 3D

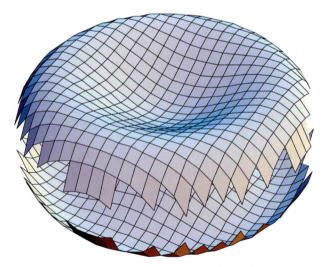

**Figure 10.1** Shape of RBC from the equation given by Evans and Fung [4]. The edge is cut open.

biconcave model of the RBC constructed using the dimensions specified in Eq. (10.1).

The well-known biconcave discoidal shape of the RBC can be easily deformed as shown in Figure 10.2 when cells are placed on polylysine-coated glass. Since negatively charged RBCs are so tightly adhered to a positively charged glass surface, the shape of the cells was changed from discoidal to that of a Mexican hat.

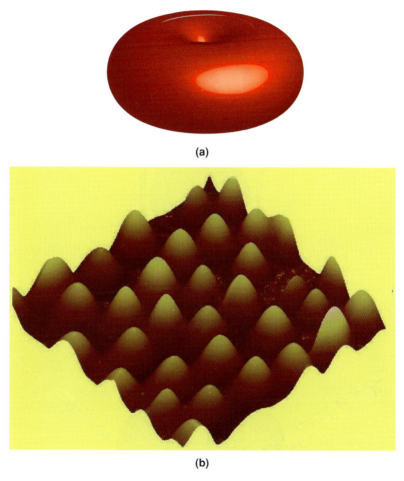

(a)

(b)

**Figure 10.2** (a) A model RBC. (b) RBCs are strongly adhered to the positively charged polylysine-coated glass. Reproduced from the study by Afrin and Ikai [4] with permission.

As has been already mentioned, the diameter of the RBC is larger than the cross-sectional diameter of capillary blood vessels. The shape of RBC is, therefore, significantly altered from the biconcave to bullet shape when it is pushed through a capillary (Figure 10.3).

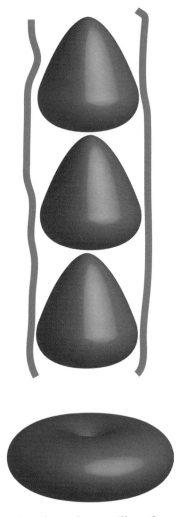

**Figure 10.3**  RBC passing through a capillary from left to right with a diameter smaller than that of RBC itself. The shape of RBC shown in this figure is often referred to as 'bullet' shaped.

## 10.2 MEMBRANE AND CYTOSKELETON

The shape of the cell is mainly determined by the intracellular cytoskeleton. The lipid bilayer membrane is very resilient, but, since it is easy to bend, it cannot maintain a defined shape without the support of the cytoskeletal structure. The easiness of bending is reflected in a low value of the bending modulus of the lipid bilayer membrane, $B$, which is in the range of $10^{-19}$ J. Though it is easy to bend the membrane, it is very difficult to extend or compress the area of the bilayer membrane with fixed number of phospholipids. This property is reflected in a low value of its lateral compressibility, $K_A$, in the range of $0.3$ N/m $(288 \pm 50$ dyn/cm) [5, 6]. This elastic constant, characterizing the resistance to area expansion or compression, is about $4 \times 10^4$ times greater than the elastic modulus for shear rigidity; therefore, the membrane behaves in a two-dimensionally incompressible liquid film under shear and elongation deformations. The low value of shear modulus implies that the lipid bilayer per se cannot retain a fixed shape. The membrane-bending modulus can be obtained also as described by Simson et al. [7].

The cytoskeleton is a network structure located immediately under the lipid bilayer and composed of several different kinds of proteins having basically a fibrous nature. In the case of RBC, the cytoskeleton is mainly made of the fibrous protein known as spectrin and actin, which are arranged in a triangulated network that lines the cytoplasmic side of the entire cell as shown in Figure 10.4. Since the cytoskeleton is almost an integral part of the RBC membrane, it is often called membrane skeleton.

The membrane skeleton is composed of roughly 33,000 hexagons per cell and actin occupies the center of each hexagon, forming the structure known as protofilament. The protofilament is tangentially associated with the lipid bilayer membrane through linker proteins such as ankyrin, but its in-plane orientation is more or less random [8].

The fact that the shape of RBC is maintained by the cytoskeleton can be proved using a ghost membrane prepared by treating RBC with the nonionic detergent, Triton, to remove phospholipid membrane. After treatment with Triton, the cell loses

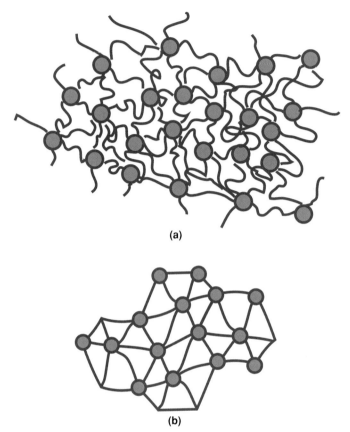

(a)

(b)

**Figure 10.4** Schematic view of RBC cytoskeleton as a hexagonal lattice of actin and spectrin. (a) in relaxed and (b) in streched states.

permeability barrier but retains the ability to change its shape between spheroidal and discoidal ones according to the change in buffer components and ATP concentration.

## 10.3 ASSOCIATION OF MEMBRANE PROTEINS WITH CYTOSKELETON

The presence or absence of association of intrinsic membrane proteins with the cytoskeletal structure is an important issue because much of the extracellular information is transferred through the cell membrane to the intracellular structures. The major candidate for this information transfer pathway is the mechanical linkage from the membrane protein to the cytoskeleton. There are several methods to probe such linkages biochemically or biophysically.

## 10.3.1 Detergent treatment

When cells are washed with one of several nonionic detergents such as Triton X-100, phospholipids in the cell membrane are dissolved in the solution together with other components in the membrane unless they are not linked to the cytoplasmic structure such as the cytoskeleton. After washing of the delipidated cellular structure, the remaining proteins can be identified by gel electrophoresis or by reaction with fluorescently labeled antibodies or lectins. When this method is applied to RBCs, a substantial amount of Band 3 is left with the spectrin-containing cytoskeleton, whereas most of glycophorin A is lost into solution. Instead of detergents, phospholipase can also be used to remove lipid membranes.

## 10.3.2 Diffusion coefficients

Membrane proteins are specifically labeled with fluorescent ligands and a part of them are photobleached by irradiating them with a laser beam of high intensity. After photobleaching, the rate of fluorescence recovery in the bleached region is monitored as a function time. The time-dependent recovery of fluorescence is due to entrance and exit of protein molecules through diffusion, having unbleached or bleached labels in and out of the prebleached area of the membrane. From the kinetic analysis of the recovery process, one can obtain 2D diffusion coefficient of specific membrane proteins. Those with unusually low diffusion coefficient are classified as being associated with the cytoskeletal components and vice versa. Fluorescence recovery rates for normal membrane indicate that more than half the labeled proteins are mobile, with a diffusion coefficient of $4 \times 10^{-15}\,\mathrm{m^2/s}$, which is in agreement with the results from other studies [9]. The diffusion coefficient for proteins in tether membrane is greater than $1.5 \times 10^{-13}\,\mathrm{m^2/s}$. This dramatic increase in diffusion coefficient indicates that extensional failure involves the uncoupling of the lipid bilayer from the membrane skeleton.

### 10.3.3 Force-curve measurement

A force curve obtained from the mechanical pulling of membrane proteins using biomembrane force probe (BFP, explained in Chapter 3) or AFM shows signs of membrane protein–cytoskeleton association. When the protein is not associated with the cytoskeleton, the force curve has a smooth feature, often characterized by an extended plateau force corresponding to the extension of membrane up to a few to tens of $\mu$m's. When those proteins known to be associated with the cytoskeleton are pulled, the force curves show several force peaks in the beginning and/or at the end of the pulling process. Such force peaks are interpreted to represent forced detachments of the membrane protein from the cytoskeleton. Afrin and Ikai showed that when glycophorin A was pulled out of the RBC surface by using a specific lectin, WGA, force curves were generally without force peaks, indicating that a substantial fraction of glycophorin A was not associated with the cytoskeleton [4]. However, when Band 3 was pulled by using concanavalin A, force curves were studded with force peaks, confirming the earlier work that more than 50% of Band 3 is associated with the cytoskeleton through the linker protein, ankyrin.

 ## 10.4 DEFORMATION OF 2D MEMBRANE

According to Landau and Lifshitz [10], 2D membrane deformation in the small deformation regime can be summarized as follows with two parameters, the area compression modulus, $K$, and the shear modulus, $\mu$. The deformation of a small area of 2D membrane $(\delta x \times \delta y)$ is shown in Figure 10.5.

$$\sigma_{xx} = K(\varepsilon_{xx} + \varepsilon_{yy}) + \mu(\varepsilon_{xx} - \varepsilon_{yy}) \qquad (10.2)$$

$$\sigma_{yy} = K(\varepsilon_{xx} + \varepsilon_{yy}) + \mu(\varepsilon_{yy} - \varepsilon_{xx}) \qquad (10.3)$$

$$\sigma_{xy} = 2\mu\varepsilon_{xy} \qquad (10.4)$$

where $\sigma_{ij} = \delta F_i / \delta x_j$ and $\varepsilon_{ij} = 1/2(\delta u_i / \delta x_j + \delta u_j / \delta x_i)$ are stress and strain in the direction of $i$th coordinate on the surface normal to the $j$th coordinate, respectively as defined in Figure 10.5.

As to the actual values of these parameters, the pipette suction method has been used to determine the value of $\mu$ since it is related to the suction pressure $P$ in the following manner.

$$P = \left(\frac{\mu}{R_P}\right)\left[\left(\frac{2L}{R_P} - 1\right) + \ln\left(\frac{2L}{R_P}\right)\right] \qquad (10.5)$$

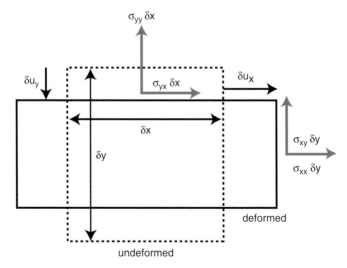

**Figure 10.5** A schematic diagram showing a deformation of a small area (dotted square) of 2 dimensional membrane $(\delta x \times \delta y)$ into a rectangle (solid line) due to various stresses in $x$ and $y$ directions. Reproduced from the study by Hénon et al. [13] with permission.

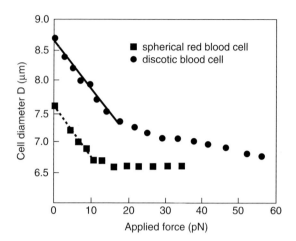

**Figure 10.6** Experimental result of RBC diameter versus applied force for a discoidal cell and a quasi-spherical cell. The slope of the linear part of the plot gives an estimate of the shear modulus. Reproduced from [13] with permission.

where $R_P$ and $L$ are the radius of the pipette and the length of the cell inside the pipette, respectively.

Values of shear modulus for RBC over the range of $6 - 9 \times 10^{-6}$ J/m² or $N/m$ have been reported by several workers [11, 12], whereas a value of $2.5 \pm 0.4 \times 10^{-6}$ J/m² was obtained by Hénon et al. [13], who used a different method as described below. Auditory hair cells are much more rigid with an effective shear stress of $1.5 \pm 0.3 \times 10^{-2}$ J/m², which is thousand times higher than that of the RBC. For a fibroblast, by using the magnetic bead method, a value of $2 - 4 \times 10^{-3}$ J/m² has been reported [14].

Boal interpreted the low value of shear modulus of the RBC membrane due to the entropic elasticity of the spectrin-based network [15]. Whereas the contour length of a spectrin tetramer that spans neighboring junctions of the network with a six-fold symmetry is 200 nm, the actual length measured *in vivo* is 75 nm on average, suggesting that spectrin tetramers are in a more or less slackened state, which is expected to show an entropic elasticity.

Hénon et al. employed the optical trap method to obtain the shear modulus of the RBC. They pulled a single cell from the two opposite ends where latex beads were attached by laser tweezers and measured the diameter of the cell, $D$, at its equatorial region as a function of the magnitude of the applied tensile force, $F$.

The result of linear-mechanics analysis gives the following relationship between $D$ and $F$, where $D_0$ is the diameter of the undeformed cell [10].

$$D = D_0 - \frac{F}{2\pi\mu}\left(1 + \left(1 - \frac{\pi}{2}\right)\frac{\mu}{K}\right) \qquad (10.6)$$

They assumed that $\mu \ll K$ and used the following equation to analyze the data as given in Figure 10.6.

$$D = D_0 - \frac{F}{2\pi\mu} \qquad (10.7)$$

The value of $\mu$ obtained by Hénon et al. was significantly lower than the those obtained using other measurements employing different methods, and the difference was attributed to small or large deformation regimes, where respective methods were applied.

## 10.5 HELFRICH THEORY OF MEMBRANE MECHANICS

In this section, a theoretical modeling of the RBC is introduced as advanced by Helfrich [16, 17]. To begin with, we review the physics of bubble formation. Suppose a spherical bubble of radius $r$ is formed from a detergent solution having the surface pressure of $\gamma$, where the air pressures outside and inside the bubble are $p_0$ and $p$, respectively. The sum of the volume and surface energy is equal to

$$\frac{4}{3}\pi r^3 \Delta p + 4\pi r^2 \gamma \qquad \text{where } \Delta p = p_0 - p, \qquad (10.8)$$

which has a minimum value when $4\pi r^2 \Delta p + 8\pi r\gamma = 0$

The result in an alternative form as shown below tells that the smaller is the bubble, the higher the internal pressure.

$$p - p_0 = \frac{2\gamma}{r} \tag{10.9}$$

When the bubble membrane is thick, the surface tension on both side of the membrane must be taken into account. The result is $p - p_0 = 4\gamma/r$.

If the bubble is not quite spherical, then we take the two principal radii of curvature, $R_1$ and $R_2$, as follows (Young-Laplace equation).

$$p - p_0 = \gamma \left( \frac{1}{R_1} + \frac{1}{R_2} \right) \tag{10.10}$$

The elastic energy of liquid membrane was formulated by Helfrich based on the consideration of elastic energy of liquid crystals as

$$V = \frac{\kappa_B}{2} \int \int (c_1 + c_2 - c_0)^2 dx dy + \kappa_G \int \int c_1 c_2 \; dx dy \tag{10.11}$$

where $c_1, c_2, c_0$, and $\kappa_B$ and $\kappa_G$ are two principal curvatures, the spontaneous curvature, bending modulus and Gaussian bending modulus of the membrane, respectively [16]. The second term relating the bending energy to Gaussian curvature is constant as long as the topology of the cell remains unchanged.

Helfrich used the above equation to explain the well-known biconcave shape of normal the RBCs [17] by first calculating the total bending energy of the cell and then minimizing it with additional constraints of constant volume and surface area. Since the average volume of 95 $\mu m^3$ is lower than the equivalent sphere for the average surface area of 135 $\mu m^2$ [18], the shape of the cell could be either oblate or prolate ellipsoid. The minimization theory of Helfrich predicted the biconcave shape of the cell as thermodynamically the most stable shape, which is in agreement with the experimental observation. In Figure 10.7, the model proposed by Deuling and Helfrich is presented. The original model shows one-fourth of the cross-section of the cell with rotational

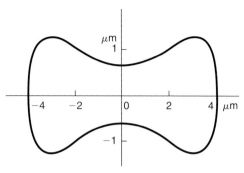

**Figure 10.7** Two-dimensional model of the RBC as proposed by Helfrich. One quarter of the equatorial cross-section is given.

symmetry around $z$-axis and mirror symmetry with respect to $x$-axis. The shape was claimed to be close to the average shape of the RBC experimentally determined by Evans and Fung [1].

 ## 10.6 CYTOPLASM AND SUBCELLULAR STRUCTURES

The cytoplasm of the eukaryotic cell is filled with the cytoskeleton, especially the intricate structures composed of microtubules, which functions as the rail for molecular motor proteins for intracellular transport of functional molecules, among other various roles. Often functional molecules are enclosed in lipid vesicles whose surface is covered with the protein called kinesin and known as the representative transport protein. The protein has two protrusions, which not only look like but also function like two legs of the human being, alternately binding and unbinding to and from the microtubule rail and moving to the specified direction on it at the expenditure of the free energy provided by ATP hydrolysis.

When the sedentary cells are imaged using an AFM, the thickness of the region of the cell not close to the nucleus is less than 1 $\mu$m, whereas the nuclear region is 3–5 $\mu$m thick. The outer rim of the region of the cell is irregularly extended, forming numerous structures called podium (pl. podia), a structure resembling or functioning as foot. Phillopodia is one such structure and their time-resolved imaging reveals a slow but rather energetic motion of the outermost parts of the live cell. Figure 10.8 shows the image of a live cell grown on a glass plate.

The sectional analysis of the image shows the change in thickness over the cell. There are other structural units in the cytoplasm such as mitochondria, lysosomes, polysomes, and a large number of lipid membranes called endoplasmic reticulum, which is actually an extension of nuclear membrane. The function of the endoplasmic reticulum is to transport proteins destined to various local areas in the cell or cell membranes and to help achieve correct folding of newly synthesized proteins, providing functions of, for example, disulfide bridge isomerase. The specialized membranous structure with the role of glycosylating proteins is called Golgi apparatus. From a biomechanical point of view, what is most interesting is the role of the cytoskeleton and endoplasmic reticulum systems as the pathways for mechanical information transfer.

The inside of the nucleus is filled with DNA and nuclear proteins. The total volume of genomic DNA which has the length of 1 m in the case of humans, is approximately $3 \times 10^{-18}\,\mathrm{m}^3$ compared with the estimated volume of the nucleus of

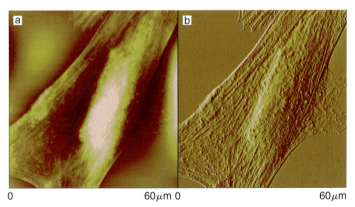

0                              60 μm 0                              60 μm

**Figure 10.8**   Image of live cells obtained using AFM. Image was taken by Ms. Salma Zohora.

$1000-8000 \times 10^{-18} \, \text{m}^3$ of typical animal cells. Although DNA itself occupies only a small portion of nucleus, it is associated with a large amount of proteins such as histones (about the same amount to DNA) and other DNA-binding proteins and the total volume of the DNA–protein complex, the chromatin, is probably 100 times larger and occupies quite a large space in the nucleus.

 ## 10.7 INDENTATION EXPERIMENT AND THE USE OF SNEDDON'S FORMULAE

### 10.7.1 Sneddon's formula

Probing the hardness of living cells of different origin and in different physiological states has been a focus of research for the last two decades [19–22]. Experimentally, an AFM probe is pushed into the cell body, and force curves such as the one in Figure 10.9 is obtained [23].

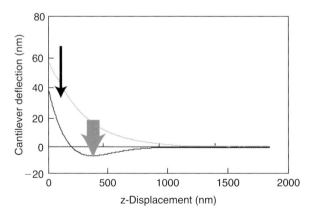

**Figure 10.9** Force curve obtained in cell-indentation experiment. The upper curve is an approach curve and the lower one is a retraction curve. The approach curve has a characteristics of indenting into a very soft material and the mismatch between the two curves as indicated by two arrows reflects viscoelastic nature of the cell in the experimental time-scale. Reproduced from the study by Afrin et al. [23] with permission.

The approach part of the force curve is analyzed according to the method of macroscopic mechanics by relating the depth of indentation to the magnitude of the applied force. The proportionality constant of the functional forms of the two variables is expressed in terms of Young's modulus and Poisson's ratio of the cell.

For the case of probe indentation into a half-plane of infinite extension, Sneddon derived analytical formulae, assuming the probes to be axi-symmetric, to be explicit [24] (a) flat cylinderical punch with a cross-sectional radius of $a$, (b) conical punch of the opening angle of $\theta$, and (c) a paraboloidal punch of a radius of curvature of $2k$, which can be used for a spherical probe of radius $R = 2k$ as long as the depth of indentation $I$ is not greater than $R$ in the following formulae.

The shapes of the three types of probe are given in Figure 10.10 together with geometrical parameters.

(a) *Flat cylindrical punch:*

$$F = \frac{4GaI}{1 - \nu} = \frac{2Ya}{(1 - \nu^2)} \qquad (10.12)$$

where $G$ is the rigidity modulus. As a reminder, $Y = 2G(1 + \nu)$.

(b) *Conical punch:*

$$F = \frac{4G\tan\theta}{\pi(1 - \nu)}I^2 = \frac{2Y\tan\theta}{\pi(1 - \nu^2)}I^2 \qquad (10.13)$$

$\theta$ is defined in Figure 10.10.

(c) *Paraboloidal punch:*

$$F = \frac{8G}{3(1 - \nu)}(2kI^3)^{1/2} = \frac{4Y\sqrt{R}}{3(1 - \nu^2)}I^{3/2} \qquad (10.14)$$

These formulae, particularly the second and the third ones, are commonly used for the analysis of indentation experiment on live cells to investigate their rigidity and its change under different physiological conditions. For example, Kawabata and his colleagues have been studying the change in rigidity [25].

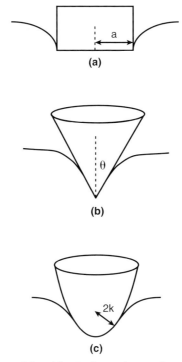

**Figure 10.10**  Types of Sneddon's formulae and corresponding probes with different shapes: (a) flat cylinder with a cross-sectional radius of $a$; (b) conical punch of an opening angle of $\theta$; (c) a paraboloidal punch of a radius curvature of $2k$ at the tip.

## 10.7.2 Correction for thin samples

When the sample is thin compared with the size of the compressing probe, the effect of hard substrate is non–negligible, and the apparent value of Young's modulus to be deduced by applying Sneddon's formulae shows gradual increase as the penetration depth increases. Such an artifact should be removed in the analysis before concluding that the hardness of the sample has a tendency to increase as the penetration becomes deeper. Though there are many attempts to remove such an artifact from compression analysis, a treatment by Dimitriadis et al. is introduced here by citing the final results from their paper [26].

First, when the sample layer is not immobilized on the substrate surface, the following relationship between the applied force and the depth of indentation was derived for a spherical probe of radius $R$. It is recommended to use a rather large spherical probe of radius of $10\,\mu\text{m}$ to avoid complication that could arise from the unknown shape and roughness of the probe.

$$F = \frac{16Y}{9} R^{1/2} I_H^{3/2} [1 + 0.884\chi + 0.781\chi^2 + 0.386\chi^3$$
$$+ 0.0048\chi^4] \tag{10.15}$$

$$\text{where } \chi = \sqrt{RI_H}/h \tag{10.16}$$

When the thin sample is immobilized on the substrate surface, the following equation, which represents a higher force for the same amount of indentation compared with the above equation, is recommended.

$$F = \frac{16Y}{9} R^{1/2} I_H^{3/2} [1 + 1.133\chi + 1.283\chi^2 + 0.769\chi^3$$
$$+ 0.0975\chi^4] \tag{10.17}$$

## 10.8 DEFORMATION MECHANICS OF A THIN PLATE

A mechanical model of cell-membrane deformation is the deformation of a thin plate as described by, for example, Landau and Lifshitz [10]. Here, we explore the nature of the flexural rigidity, $D$, of a two dimensional thin plate, which is expressed in terms of Young's modulus and Poisson's ratio of the material that makes up the plate and its thickness $h$.

$$D = \frac{Yh^3}{12(1 - \nu^2)} \tag{10.18}$$

According to Landau and Lifshitz [10], the free energy of the deformation $\phi$ of linearly elastic material is, in general, given by the following equation (10.19), where the second and the third term on the right are, respectively, related to extension/compression and the shear deformation, in the sense that, $u_i = x'_i - x_i$ for ($i = 1, 2, 3$) is the tensile deformation of a small volume element in three axes, 1, 2, and 3, from the original position of $x_i$ to $x'_i$. In addition, $u_{ik} = (1/2)(\partial u_i/\partial x_k + \partial u_k/\partial x_i)$ and when a term has two identical suffices such as $u_{ii}$, it means summation over three axes such that, $u_{ii} = u_{11} + u_{22} + u_{33}$. (This notation is known as Einsteins's general summation rule.)

$$\phi = \phi_0 + \frac{1}{2}\lambda u_{ii}^2 + G u_{ik}^2 \quad u_{ii} = u_{11} + u_{22} + u_{33} \quad \text{and}$$

$$u_{ik} = \frac{1}{2}\left(\frac{\partial u_i}{\partial x_k} + \frac{\partial u_k}{\partial x_i}\right) \tag{10.19}$$

where $\lambda$ and $G$ are called Lamé coefficients. When there is no volume change, the second term on the right is zero and the deformation is a 'pure shear'. $G$ is rigidity modulus and equal to $Y/[2(1 + \nu)]$ and $\lambda = 2G\nu/(1 - 2\nu) = Y\nu/[(1 - 2\nu)(1 + \nu)]$.

Thus, in terms of Young's modulus and Poisson's ratio, the expression for the free energy is

$$\phi = \frac{Y}{2(1 + \nu)}\left(u_{ik}^2 + \frac{\nu}{1 - 2\nu}u_{ii}^2\right) \tag{10.20}$$

In the case of a bent plate under an external force of $P$, $u_{ik}$ and $u_{ii}$ are given in the following form in the Cartesian coordinates of $x, y$, and $z$, with a deflection of the plate from the neutral line equal to $\zeta$ as shown in Figure 10.11.

$$u_{xy} = -z\frac{\partial^2 \zeta}{\partial x \partial y} \tag{10.21}$$

$$u_{zz} = z\frac{\nu}{1 - \nu}\left(\frac{\partial^2 \zeta}{\partial x^2} + \frac{\partial^2 \zeta}{\partial y^2}\right) \tag{10.22}$$

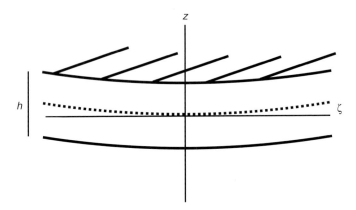

**Figure 10.11** A plate of thickness $h$ is bent so that its deflection from the horizontal line is equal to $\zeta$ in $x, y$, and $z$ coordinate system.

By substituting the above expression in $\phi$,

$$\phi = z^2 \frac{Y}{1+\nu} \left\{ \frac{1}{2(1-\nu)} \left( \frac{\partial^2 \zeta}{\partial x^2} + \frac{\partial^2 \zeta}{\partial y^2} \right)^2 \right.$$

$$\left. + \left[ \left( \frac{\partial^2 \zeta}{\partial x \partial y} \right)^2 - \frac{\partial^2 \zeta}{\partial x^2} \frac{\partial^2 \zeta}{\partial y^2} \right] \right\} \tag{10.23}$$

The total free energy of the plate is obtained by integrating over the entire volume of the plate. The integration over $z$ ranges from $-\frac{1}{2}h$ to $+\frac{1}{2}h$, where $h$ is the thickness of the plate and that for $x$ and $y$ is over the surface of the plate. The total free energy of the deformed plate, $\phi_{pl} = \int \phi dV$, is then,

$$\phi_{pl} = \frac{Y h^3}{24(1-\nu^2)} \int \int \left[ \left( \frac{\partial^2 \zeta}{\partial x^2} + \frac{\partial^2 \zeta}{\partial y^2} \right)^2 + 2(1-\nu) \right.$$

$$\left. \left\{ \left( \frac{\partial^2 \zeta}{\partial x \partial y} \right)^2 - \frac{\partial^2 \zeta}{\partial x^2} \frac{\partial^2 \zeta}{\partial y^2} \right\} \right] dx dy$$

$$\tag{10.24}$$

By equating the variation in $\phi_{pl}$ to the work done by an external force $P$ when the position on the plate is displaced at a distance $\zeta$ is $\int F\delta\zeta\mathrm{d}f = 0$.

$$\delta\phi_{pl} - \int F\delta\zeta\mathrm{d}f = 0 \qquad (10.25)$$

For the following integral to be equal to zero, since the variation in the integral, $\delta\zeta$, is arbitrary, $D\Delta^2\zeta - F = 0$. This is called the equation of equilibrium for a plate bent by external force $F$ acting on it.

$$D\Delta^2\zeta - F = 0 \qquad (10.26)$$

where

$$D = \frac{Yh^3}{12(1 - \nu^2)} \qquad (10.27)$$

is called 'flexural rigidity' of the plate. This definition of $D$ is found in the literature on physical treatment of biomembranes.

## Bibliography

[1] Evans, E. and Fung, Y. C. (1972), Improved measurements of the erythrocyte geometry, *Microvasc. Res.*, 4, 335–347.

[2] McMillan, D. E., Mitchell, T. P., and Utterback, N. G. (1986), Deformational strain energy and erythrocyte shape, *J. Biomech.*, 19, 275–286.

[3] Nakao, M. (2002), New insights into regulation of erythrocyte shape, *Curr. Opin. Hematol.*, 9, 127–132.

[4] Afrin, R. and Ikai, A. (2006), Force profiles of protein pulling with or without cytoskeletal links studied by AFM, *Biochem. Biophys. Res. Commun.*, 348, 238–244.

[5] Evans, E. A. (1983), Bending elastic modulus of red blood cell membrane derived from buckling instability in micropipet aspiration tests, *Biophys. J.*, 43, 27–30.

[6] Evans, E., and Yeung, A., (1989), Apparent viscosity and cortical tension of blood granulocytes determined by micropipet aspiration, *Biophys. J.*, 56, 151–160.

[7] Simson, R., Wallraff, E., Faix, J., Niewöhner, J., Gerisch, G., and Sackmann, E. (1998), Membrane Bending Modulus and Adhesion Energy of Wild-Type and Mutant Cells of Dictyostelium Lacking Talin or Cortexillins, *Biophys. J.*, 74, 514–522.

[8] Picart, C., Dalhaimer, P., and Discher, D. E. (2000), Actin protofilament orientation in deformation of the erythrocyte membrane skeleton, *Biophys. J.*, 79, 2987–3000.

[9] Berk, D. A. and Hochmuth, R. M. (1992), Lateral mobility of integral proteins in red blood cells tethers, *Biophys. J.*, 61, 9–18.

[10] Landau, L. D. and Lifshitz, E. M. (1986), 'Theory of Elasticity', 3rd English edition, Butterworth-Heinemann, Oxford, UK.

[11] Evans, E. A., Waugh, R., and Melnik, L. (1976), Elastic area compressibility modulus of red cell membrane, *Biophys. J.*, 16, 585–595.

[12] Waugh, R. and Evans, E. A. (1979), Thermoelasticity of red blood cells membrane, *Biophys. J.*, 26, 115–131.

[13] Hénon, S., Lenormand, G., Richert, A., and Gallet, F. (1999), A new determination fo the shear modulus of the human erythrocyte membrane using optical tweezers, *Biophys. J.*, 76, 1145–1151.

[14] Bausch, A. R., Ziemann, F., Boulbitch, A. A., Jacobson, K., and Sackmann, E. (1998), Local measurements of viscoelastic parameters of adherent cell surfaces by magnetic bead micro-rheometry, *Biophys. J.*, 75, 2038–2049.

[15] Boal, D. (2002), 'Mechanics of the Cell', Chapter 3, 59–95. Cambridge University Press, Cambridge, UK.

[16] Helfrich, W. (1973), Elastic properties of lipid bilayers: theory and possible experiments, *Z. Naturforsch [C].*, 28, 693–703.

[17] Deuling, H. J. and Helfrich, W. (1976), Red blood cell shapes as explained on the basis of curvature elasticity, *Biophys. J.*, 16, 861–868.

[18] Fung, Y. C. (1993), 'Biomechanics: Mechanical Properties of Living Tissues', Springer, New York, NY.

[19] Rotsch, C. and Radmacher, M. (2000), Drug-induced changes of cytoskeletal structure and mechanics in fibroblasts: an atomic force microscopy study, *Biophys. J.*, 78, 520–535.

[20] Radmacher, M. (2002), Measuring the elastic properties of living cells by the atomic force microscope, *Methods Cell Biol.*, 68, 67–90.

[21] A-Hassan, E., Heinz, W. F., Antonik, M. D., D'Costa, N. P., Nageswaran, S., Schoenenberger, C. A. et al. (1998), Relative microelastic mapping of living cells by atomic force microscopy, *Biophys. J.*, 74, 1564–1578.

[22] Tao, N. J., Lindsay, S. M., and Lees, S. (1992), Measuring the microelastic properties of biological material, *Biophys. J.*, 63, 1165–1169.

[23] Afrin, R., Yamada, T., and Ikai, A. (2004), Analysis of force curves obtained on the live cell membrane using chemically modified AFM probes, *Ultramicroscopy*, 100, 187–195.

[24] Sneddon, I. N. (1965), The relation between load and penetration in the axisymmetric Boussinesq problem for a punch of arbitrary profile, *Int. J. Eng. Sci.* 3, 47–57.

[25] Mizutani, T., Haga, H., and Kawabata, K. (2007), Development of a device to stretch tissue-like materials and to measure their mechanical properties by canning probe microscopy, *Acta Biomater.*, Jan 22 (epub ahead of print).

[26] Dimitriadis, E. K., Horkay, F., Maresca, J., Kachar, B., and Chadwick, R. S. (2002), Determination of elastic moduli of thin layers of soft material using the atomic force microscope, *Biophys. J.*, 82, 2798–2810.

CHAPTER ELEVEN

# MANIPULATION AT THE MOLECULAR LEVEL

## Contents

Knowing the mechanical properties of proteins, DNA, and organelles and cells we can manipulate these biological molecules and structures by applying force externally. As already explained in earlier chapters, proteins, DNA, and RNA may be mechanically stretched and refolded, and the cells may be poked to test their responses against mechanical stresses.

## 11.1 PROSPECTS FOR USEFUL APPLICATIONS OF NANOMECHANICS

There seems to be an exciting possibility in the use of mechanical force to accelerate the rate of otherwise very slow events in nature such as unzipping of double helical DNA, unfolding of globular proteins, unbinding of tightly associated ligand–receptor pairs, uprooting intrinsic membrane proteins from the lipid bilayer,

uncoiling nucleosomal DNA from histone cores, and making a
hole in the cell membrane, among other possibilities. Using such
technologies, we will be able to operate on single molecules
of DNA and proteins, cell membranes, cytoplasm, and finally
perform live-cell surgeries.

## 11.2 CELL SURGERY

The obvious target of application of nano-biomechanics is the
cellular-level surgery in a sense that a microsurgical knife comes in
contact with a specific local area on a living cell and manipulates
the activity and localization of membrane proteins or penetrates
the cell and operates on the intracellular structures to correct any
of their defects. It will recover a small amount of proteins from the
cell membrane or mitochondria, and other subcellular structures
from the inside of the nucleus. By operating on the genetic material in the nucleus and mitochondria, the cell doctor can change
the properties of the offsprings of the operated cells. If the cell is an
embryonic stem cell, the tissues and organs to be reproduced from
it will have its altered characteristics. Thus, there is an enormous
potential for cellular-level surgery in the future.

The first problem to be solved is how to bore a hole on the
cell membrane without a considerable injury to the cell. Little is
known about the recovery process of the injured cell or about the
critical size of the hole for recovery of the cell. Now, experiment
conducted in several laboratories have led to the accumulation of
seminal knowledge about these processes [1].

## 11.3 CHROMOSOMAL SURGERY AND GENE
##          MANIPULATION

Presently, since it has been possible to insert plasmid DNA into a
live cell by using AFM tips [2, 3], there is a good chance of using
nanotechnology to change the genetic information sequestered in

the genomic DNA, though there are yet many conceivable barriers to perform such operation routinely. The following steps must be cleared before we are able to do genome surgery on live cells.

1. The genomic DNA must be pulled out of the nucleus in an intact form and stretched on an operational table.

2. The part on the DNA to be operated must be clearly indicated on the linearly stretched DNA.

3. The part of the DNA that contains defective base sequences should be cut and removed from the mother gene.

4. A replacement DNA segment must be brought to the position and ligated to the DNA just operated.

5. The genomic DNA must be put back in the nucleus without any damage to the rest of the base sequence.

6. The cell that received the operated DNA must survive and proliferate, without any damages being inflicted on its differentiation potential.

It will take a long time before all the requirements listed above are to be fulfilled, but when they are fulfilled, the possibility of rectifying the genetic problems in every single cell may find useful applications. The ethic problem of tinkering with the genetic material, which is the product of long evolutionary history and a gift from your parents, should be considered once again before the technology becomes imminent. The least what we could say is that the genome we have today has been shaped by random mutations in the base sequence and by spreading of particular mutations according to the law of natural selection.

## 11.4 TISSUE SURGERY

Operations at the tissue level using nanomechanical technology have been tried in knee surgery. A small AFM-based mechanical sensor was inserted in the knee joint to measure the mechanical properties of hardening tissue [4]. To study osteoarthritis at early stages, a miniaturized and integrated scanning force microscope was built to fit into a standard arthroscopic device as shown in Figure 11.1. This instrument allows orthopedic surgeons to measure the mechanical properties of articular cartilage at the

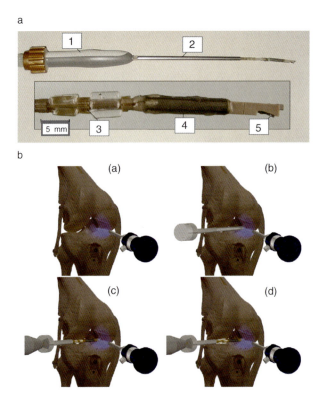

**Figure 11.1** (A) Picture of the scanning force arthroscope (SFA). The inset exhibits a magnified image of the distal end of the instrument containing from left to right, the scanning stage (3), the tube scanner (4), and the IT AFM sensor (5). (B) (bottom) The procedure employed for safe positioning and stabilization of the SFA: (a) Visual inspection of the knee by means of an optical arthroscope. (b) Creation of an entry access for the instrument. (c) Insertion of the SFA. (d) Positioning and stabilization of the instrument by inflating the balloons. Reproduced from the study by Imer et al. [4] with permission.

nanometer and micrometer scales in vivo during a standard arthroscopy. Challenging work as the one just described will follow in various fields of medicine to allow surgeons to monitor and operate on a small part of the injured tissue at a minimal invasion level.

## 11.5 LIPOSOMAL TECHNOLOGY

Liposomes are small bags made of phospholipid bilayers. There are single-layered liposomes and multilayered liposomes. Single-layered liposomes are used as a model of the cell. From our point of interest in this study, incorporation of proteins or protein-synthesizing machineries into a liposome stands out as a budding trial for construction of an artificial cell. Incorporation of actin or tubulin molecules inside a single-layered liposome allowing for the polymerization of monomers, was used as a model for cytoskeleton structure. With the growth of fibrous structures inside the liposomer, the shape of the liposome gradually elongated, almost mimicking the structural changes occurring in a moving cell [5].

Nomura et al. showed that the protein-synthesizing machineries incorporated inside a liposome produced a particular protein [6]. Several liposomes can be functionally connected through hollow lipid nanotubes formed as extensions of a single phospholipid bilayer. Akiyoshi and colleagues showed that mixing of cholesterol with phospholipids as the starting material for liposome formation helped formation of tubular structures, which could be used as a pipe to move fluorescence dyes from one liposome to another by diffusion [7, 8].

## 11.6 DRUG DELIVERY

Drug delivery is a strongly focused application of nanobiology. Drugs are now designed such that it is transported to the specific pathological sites where the drug is required for curing particular diseases. Drug itself cannot target, for example, a cancerous tissue, but when the drug is tagged with specific label that recognizes particular types of cancer cells and is efficiently internalized, the drug delivery is efficient and economical and alleviates many side effects in the patient. This basic concept of intelligent drug delivery has been explored with intense efforts all over the world. In most cases, a specific drug is encapsulated in small containers whose outer surface is labeled with a specially prepared tag(s) that has a specific affinity for marker molecules on the cancer cell surface. Often the tag is antibodies raised against membrane proteins expressed solely on cancer cells.

Where does the mechanics come in this well-formulated system of drug delivery? Park et al. reported an interesting case of transdermal drug delivery by creating holes in the skin by preparing an array of microneedles made of biodegradable polymers [9]. They found that the mechanical properties and biocompatibility of the needles made of the polymer were satisfactory for the purpose. Polylactic acid (PLA), polyglycolic acid (PGA), and the copolymers (PLGA) of the two, which were used because of their established biocompatibility, had high mechanical strength of 1 GPa (for low-MW PLGA) to 10 GPa (PGA) in terms of Young's modulus. The force during the application for creating holes on the skin was in the order of 1/10 N, and the efficiency of drug delivery was increased by three orders. The force necessary to deliver drugs into the cadaver skin was well under the yield force of the polymer needles. In the case of drug delivery into individual living cells, this type of approach is possible and should be investigated.

## 11.7 DNA AND RNA RECOVERY FROM THE CHROMOSOME AND THE CELL

For monitoring the biochemical status of live cells in culture, it is necessary to analyze the time-dependent change of the cellular materials without either killing or damaging the cell too seriously. To begin with this type of approach, experiments on DNA extraction from chromosomes and mRNA extraction from cytoplasm are introduced in this section. The method to pull out membrane proteins from live-cells as previously explained is also along this line of live-cell monitoring.

AFM can be used to extract DNA from chromosomes or mRNA from the cytoplasm. Xu and Ikai demonstrated that a single copy of genomic DNA could be extracted from an isolated mouse chromosome and be PCR amplified and sequenced [10]. A piece of chromosome was first imaged with a probe that was amino–silanized at pH 10, in which case the amino groups were not protonated. After imaging, the probe was brought to a selected site on the chromosome and the pH of the solution was lowered to 7, in which case the probe was fully protonated. Under this condition where the electrostatic interaction between the positively charged probe versus negatively charged DNA was maximized, the probe was pushed into the chromosome and then pulled out together with some of the DNA segments. The force mode of AFM operation recorded the pull-out process of DNA as a prolonged downward deflection of the cantilever as shown in Figure 11.2.

The probe presumably having extracted DNA was used as the source for PCR amplification, and the amplified DNA was used for the confirmation of the extraction position by fluorescence in-situ hybridization (FISH) method. The amplified DNA was then sequenced.

Uehara et al. showed that mRNA can be extracted from the cytoplasm of live cells as it adsorbed to an AFM probe which was inserted into the cell by the application of a strong force [11, 12]. After the recovery of the AFM probe they placed it in a test tube containing the ingredients for RT-PCR amplification and then performed an ordinary PCR amplification. They successfully detected the presence of the mRNA for the household protein, $\beta$-actin, in 173 cases out of a total of 176 probes inserted in live cells (success rate $= 97\%$). It was possible to show the localization of $\beta$-actin mRNA within a cell and also the

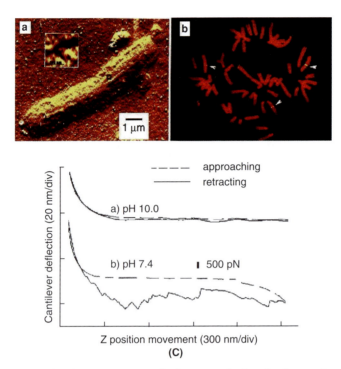

**Figure 11.2** DNA was extracted from an isolated piece of mouse chromosome. The force curves obtained at pH 10 showed no downward deflection on retraction from the chromosome, but a similar operation at pH 7 was characterized by the force curves showing prolonged downward deflection, signifying pulling out of DNA. Reproduced from [10] with permission.

dependence on the physiological stage of the cell as described in Figure 11.3. In resting cells, $\beta$-actin mRNAs were localized closer to the nucleus, whereas in an activated state by a supply of nutrients, finite levels of $\beta$-actin mRNA were detected away from the nucleus but only in the frontal part with respect to the cell locomotion. This method can be developed for determining a precise intracellular distribution of particular mRNAs as a function of time because the method does not result in lethalrty to the cells.

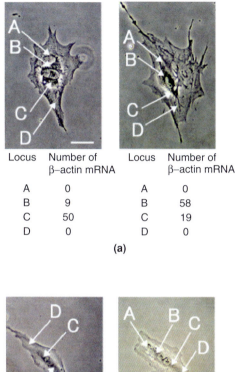

| Locus | Number of β–actin mRNA | Locus | Number of β–actin mRNA |
|-------|------------------------|-------|------------------------|
| A | 0 | A | 0 |
| B | 9 | B | 58 |
| C | 50 | C | 19 |
| D | 0 | D | 0 |

(a)

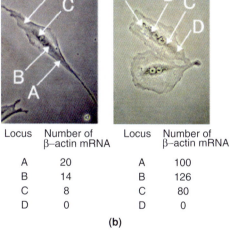

| Locus | Number of β–actin mRNA | Locus | Number of β–actin mRNA |
|-------|------------------------|-------|------------------------|
| A | 20 | A | 100 |
| B | 14 | B | 126 |
| C | 8 | C | 80 |
| D | 0 | D | 0 |

(b)

**Figure 11.3** mRNAs for $\beta$-actin were extracted from the cytoplasm of different loci of individual cells as marked by A, B, C, and D, each time by changing AFM probes. The mRNAs adsorbed on the probes were then amplified using RT-PCR and PCR methods. Cells in (a) are resting cells and those in (b) are activated cells obtained by the addition of calf serum to the culture medium. Reproduced from the study by Uehara et al. [12] with permission.

# Bibliography

[1] Zohora, U. S. (2007), Atomic Force Microscope (AFM) studies for single cell manipulation MS thesis. Tokyo Institute of Technology, Tokyo, Japan.

[2] Han, S., Nakamura, C., Obataya, I., Nakamura, N., and Miyake, J. (2005), Gene expression using an ultrathin needle enabling accurate displacement and low invasiveness, *Biochem. Biophys. Res. Commun.*, 332, 633–639.

[3] Obataya, I., Nakamura, C., Han, S., Nakamura, N., and Miyake, J. (2005), Nanoscale operation of a living cell using an atomic force microscope with a nanoneedle, *Nano Lett.*, 5, 27–30.

[4] Imer, R., Akiyama, T., de Rooij, N. F., Stolz, M., Aebi, U., Kilger, R. et al. (2007), In situ measurements of human articular cartilage stiffness by means of a scanning force microscope, *J. Phys: Conf. Ser.*, 61, 467–471.

[5] Inaba, T., Ishijima, A., Honda, M., Nomura, F., Takiguchi, K., and Hotani, H. (2005), Formation and maintenance of tubular membrane projections require mechanical force, but their elongation and shortening do not require additional force, *J. Mol. Biol.*, 348, 325–333.

[6] Nomura, S. M., Tsumoto, K., Hamada, T., Akiyoshi, K., Nakatani, Y., and Yoshikawa, K. (2003), Gene expression within cell-sized lipid vesicles, *Chembiochem.*, 4, 1172–1175.

[7] Akiyoshi, K., Itaya, A., Nomura, S. M., Ono, N., and Yoshikawa, K. (2003), Induction of neuron-like tubes and liposome networks by cooperative effect of gangliosides and phospholipids, *FEBS Lett.*, 534, 33–38.

[8] Nomura, S. M., Mizutani, Y., Kurita, K., Watanabe, A., and Akiyoshi, K. (2005), Changes in the morphology of cell-size liposomes in the presence of cholesterol: formation of neuron-like tubes and liposome networks, *Biochim. Biophys. Acta*, 1669, 164–169.

[9] Park, J. H., Allen, M. G., and Prausnitz, M. R. (2005), Biodegradable polymer microneedles: fabrication, mechanics and transdermal drug delivery, *J. Control. Release*, 104, 51–66.

[10] Xu, X. M. and Ikai, A. (1998), Retrieval and amplification of single-copy genomic DNA from a nanometer region

of chromosomes: a new and potential application of atomic force microscopy in genomic research, *Biochem. Biophys. Res. Commun.*, 248, 744–748.

[11] Osada, T., Uehara, H., Kim, H., and Ikai, A. (2003), mRNA analysis of single living cells, *J. Nanobiotechnol.*, 1, 2.

[12] Uehara, H., Osada, T., and Ikai, A., (2004), Quantitative measurement of mRNA at different loci within an individual living cell, *Ultramicroscopy*, 100, 197–201.

## CHAPTER TWELVE

# Finite Element Analysis of Microscopic Biological Structures

S. Kasas[1,2], T. Gmur[3] and G. Dietler[1]

## Contents

1 Laboratoire de physique de la matière vivante, Ecole Polytechnique Fédérale de Lausanne, 1015 Lausanne, Switzerland
2 Département de Biologie Cellulaire et de Morphologie, Université de Lausanne, Bugnion 9, 1005 Lausanne, Switzerland
3 Laboratoire de mécanique appliquée et d'analyse de fiabilité, Ecole Polytechnique Fédérale de Lausanne, 1015 Lausanne, Switzerland

## 12.1 INTRODUCTION

Nowadays, finite-element procedures are being implemented in almost all engineering disciplines, thousands of structural, thermal, fluid, electrical and electromagnetic models being solved daily using this method. This success lies in the capability of the method to analyze steady, transient, linear, and nonlinear problems that may embrace more than one type of physical phenomenon, such as interaction between a fluid and solid or one between heat and an electromagnetic field. On a somewhat smaller scale, the research community has also adopted finite-element analysis techniques to predict or validate its measurements. Biologists and physicians have only lately begun to realize the tremendous potential of the method. But the existing gulf between engineering and biomedical sciences is rapidly narrowing owing to advancements in the speed of microprocessors and to the availability of affordable computer hardware and finite element software. The most obvious biomedical application of finite element modeling is in the musculoskeletal system. However, an increasing number of analyses are being conducted on microbiological structures. After a brief historical perspective of the principle and a description of the methodology, this chapter will focus on the microbiomedical applications of this tool.

## 12.2 A BRIEF HISTORY OF THE FINITE ELEMENT METHOD

The behavior of simple entities, such as rods or beams, can be described fairly easily on the basis of elementary principles. Indeed, the relationship that describes the forces acting on these entities and the terminal displacements thereby a brought about is almost trivial. A large number of such entities can be assembled into a complex structure. An analysis based on the displacements of such complex structures was first suggested by Louis Navier in 1826. In 1910, Richardson adopted a similar approach (the finite difference method) to approximate the plain stress of a masonry dam. He recruited boys from the local high-school to make the necessary numerical calculations and referred to these youths as his 'computers'. They were paid according to the number of coordinate point calculations made and the number of digits used. If his 'computers' made errors, they were not paid! Thirteen years later, in 1923, Courant subdivided the problem into triangles [1]. In 1943, he published his completed findings, which represent the first published use of a prototype triangular element to solve torsion problems. A major breakthrough came in 1956 when Turner demonstrated how complex in-plane plate problems could be represented by finite triangular elements. In 1960, Clough [2] fully accounted mathematically for the success of the subdivision of the problem into elements. He demonstrated how the approximate solution converges to the exact mathematical solution as the size of the elements decreases. During the 1960s, more sophisticated finite elements were developed. In 1963, Melosh [3] realized that the finite element method could be extended to field problems by implementing variational methods. His paper was an important contribution in that it led to a much broader application of the finite element method, which was extended to include numerous steady-state and transient field problems. In the early 1960s, the integrated circuit was developed, and in 1970, Intel invented the first microprocessor. During this period, several commercial finite element packages (Sadsam 1960, Nastran 1965, Ansys 1970, Marc 1972) were released in the market, which opened up the method to the scientific community at large. Since this time, computational power has increased and became cheaper, thereby permitting the running of finite element programs even on microcomputers. Review articles describing the history of the finite element method have been written by Samuelson and Zienkiewicz [4] and Clough [5]. For a more mathematical treatment of the principle, the reader is referred to the publication by Thomée [6].

## 12.3 THE FINITE ELEMENT METHOD

Several physical situations can be described using differential equations according to their initial and boundary conditions. These equations are derived by applying fundamental physical laws such as those describing the conservation of mass, force, or energy. Sometimes, these governing differential equations can be solved exactly by analytical means. However, in many practical situations, exact solutions cannot be obtained, owing either to the complex nature of the governing differential equations or to difficulties arising from the initial and boundary conditions. To solve these types of problems, one has to rely on numerical approximation techniques. In contrast to analytical solutions, which describe the exact behavior of the system at any point within its volume, numerical approximation techniques yield an exact solution only at discrete points, which are referred to as nodes. The two most commonly used numerical procedures are the finite difference and the finite element methods. In the former case, the differential equations are written for each node of the system and the derivatives are replaced by difference equations. The solution is achieved by solving the set of simultaneous linear equations that describe the system. Although finite difference methods are fairly intuitive and easy to comprehend, they are not readily applicable to structures that are characterized either by complex geometries, by anisotropic material properties, or by complex boundary conditions. The finite element method similarly divides the system into nodes (and elements). But it implements an integral formulation to generate the system of algebraic equations. Furthermore, an approximate continuous function is assumed to represent the solution for each element. The final solution is obtained by assembling the individual solutions for each element, thereby ensuring continuity of the solution between these.

The finite element method thus comprises the following basic steps,

- *Definition of the geometric shape and of the material properties of the body*

- Discretization of the body into nodes and elements (meshing)

- Assumption of a shape function to represent the physical behavior of the elements.

- Development of equations to describe an element.
- Assemblage of the elements to describe the overall problem.
- *Application of the boundary conditions.*
- Simultaneous solution of a set of linear or nonlinear algebraic equations to yield nodal results.
- *Display of data (postprocessing).*

Details about these different steps can be found in almost any book dealing with the subject [7, 8]. If one plans to use a commercially (or freely) available finite element package rather than to develop one's own computer programs, only the italicized steps need be completed. An example of a finite element simulation of the indentation of an erythrocyte using an atomic force microscope is shown in Figure 12.1.

Only the italicized steps of the procedure are shown. But the apparent simplicity of the remaining process should not beguile the user into believing that a finite element analysis can be performed without a thorough understanding of the physical phenomena underlying the problem or a good comprehension of the manner in which it is solved using the programs. Furthermore, the highly-specialized finite element software must be mastered. For these reasons, the task of analysis in most engineering companies is given to a specialized member of the team, who is experienced in this type of modeling. The pitfalls to be avoided and the limitations of the method, as well as the approach that must be pursued to generate reliable finite element models, are well described by Adams and Askenazi [9].

## 12.4 APPLICATION OF THE FINITE ELEMENT METHOD TO MICROBIOLOGICAL SAMPLES

We will now focus on studies in which the finite element method has been used to calculate the mechanical properties or to predict

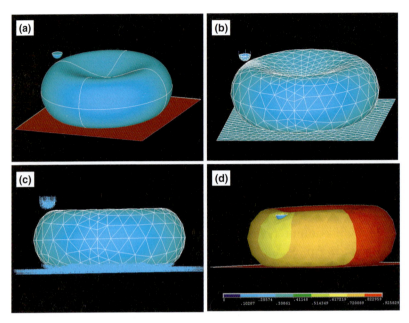

**Figure 12.1**   The main steps that build up a typical finite element analysis (simulation completed with the commercially available software ANSYS® 10.0): (a) definition of the geometry, (b) discretization of the tip, the tip supporting springs, the sample and the substrate into nodes and elements (meshing), (c) Definition of the boundary conditions (small triangles indicate spots where boundary conditions have been applied : the tip motion was restricted to the z direction only, whereas the bottom of the erythrocyte as well as the substrate were blocked in the x and y directions), (d) display of the global displacement vector in false colors.

the behavior of complex nano- and microscale biological samples. Necessarily, the published reports initially furnish a description of, or cite, the particular method that is employed to measure the mechanical properties of the microscopic sample. A comprehensive review of the different methodological tools utilized (*e.g.*, atomic force microscopy and laser tweezering) has been written by Van Vliet and Bao [10]. Here, we will not discuss the simulation of electromagnetic or thermal properties, which is poorly represented in the literature.

## 12.4.1 Proteins

The simulation of single proteins by the finite element method is a relatively new area, which is likely to develop rapidly in the near future. The reason for this is that the 'molecular dynamics' technique currently employed for such purposes cannot model the behavior of large proteins over the entire span of their physiological functioning time, which may be in the order of milliseconds. Furthermore, models developed by the molecular dynamics technique are ill-suited to simulate the deformation of large proteins under loading conditions. Finite element approximations are thus a valuable alternative in many instances such as those described below. Together with actin, spectrin is a major constituent of the cytoskeleton in human erythrocytes. During its lifespan of 120 days, each 7.5-micrometer-diameter corpuscle undergoes thousands of severe elastic deformations during its passage through narrow capillaries whose inner diameter can be as small as 3 micrometers. Consequently, the characteristically biconcave erythrocyte is periodically transformed into a bullet-shaped form [11]. But this physiological phenomenon is only one of several reasons why the deformation of human red blood cells has long been a topic of considerable scientific interest [12]. Another incentive has been that the progression of certain inherited (sickle cell anemia) and parasitic diseases (malaria) is associated with the deformation characteristics of the red blood cell. The mechanical properties of an erythrocyte depend largely on its cytoskeleton. And this is the reason why the elastic properties of spectrin and of its network have been widely investigated.

In 1997, Hansen et al. developed a finite element model which incorporated the intrinsic elastic properties of spectrin and the geometric organization of the cytoskeletal network in the red blood cell [13]. In this model of the network, the spectrin molecules were represented by springs and elastic rod elements, whereas the nodes corresponded to the protein junctions in the cytoskeleton. This model was used to compute the macroscopic mechanical properties of the erythrocyte plasmalemma. Comprehensive reviews of the structure and function of spectrin have been written by De Matteis and Morrow [14]. The mechanical properties of spectrin networks are described in detail in Ref. [12].

Recently, a sophisticated finite element model has been developed to simulate the function of bacterial mechano-sensitive channels [15]. In this model, the transmembrane proteins are represented as elastic rods and the lipid bilayer within which they are sealed, as an elastic sheet. The forces acting on this model were derived from molecular dynamics simulations. Although such finite element modeling permits a simulation of the behavior of large molecular structures over long periods of time, the approach is poorly exploited.

A few examples are given below to illustrate the finite element modeling of large protein assemblies, namely, structures whose lengths range from tens of nanometers to several micrometers. These large macromolecular structures are better suited than smaller ones to the finite element methodology, since their size renders them free of quantum-mechanical effects and amenable to modeling by continuum mechanics approximations. A well-known and representative member of this family of large protein structures are the microtubules. These structures, which constitute a major component of the cytoskeleton, play an essential role in many fundamental cellular processes. They furnish mechanical stability to the cell and act as tracts for the transport of vesicles, organelles, and chromosomes by motor proteins. These functions reflect the peculiar structure of the microtubules, which has been fairly well elucidated [16], and their mechanical properties, which are still under debate. The mechanical strength of microtubules has been measured since 1979 using various techniques, which have yielded diverse values ranging from 1 MPa [17] to 7 GPa [18]. This large discrepancy probably reflects a change in our conception of these structures. Untill 2002, microtubules were conceived in the models to be simple, homogeneous, isotropic 'tubes'. But in 2002, measurement of their Young's modulus and their shear modulus by atomic force microscopy [19] permitted the construction of more reliable models. Kis et al. used their values to construct a finite element model of microtubules in which the protofilaments and their interconnections were represented as beams, whose mechanical

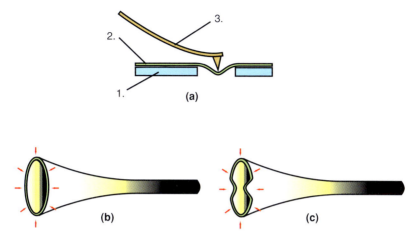

**Figure 12.2**    (a) Setup for the measurement of the mechanical characteristics of microtubules: 1) Substrate with holes, 2) microtubule, 3)
cantilever and tip of the atomic force microscope. (b) Two oscillation
modes of the microtubules as revealed by finite element analysis.

properties were adjusted to suit the experimental data. This model
was eventually used to validate the experimental procedure, to
determine its boundary conditions, and to predict several characteristics of microtubules such their oscillation (see Figure 12.2),
their susceptibility to the removal of single tubulin subunits,
and the mechanical properties of exotic microtubules [20, 21].
More recently, Schaap et al. have used finite element modeling
to validate atomic force microscopy measurements pertaining to
microtubules [22].

## 12.4.2 Axonemata and cilia

An axoneme is the active axial cytoskeletal structure of eukaryotic
flagella and cilia. It consists of a central pair of singlet microtubules
surrounded by nine doublet microtubules as shown in Figure 12.3.

The doublets are cross-linked at various points to form a
bundle; they are also linked by ciliary dynein molecular motors.

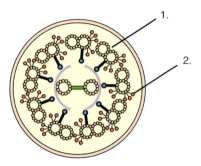

**Figure 12.3**  Schematics of an axoneme's section formed by 9 doublets plus 2 central individual microtubules: (1) tubulin doublets, (2) dynein arms.

Since the microtubules are cross-linked and basally affixed, the formation of dynein links causes a bending of the filaments. This bending induces flagella and cilia to undulate, thereby permitting the motion of spermatozoa, a cleaning of the airways and the establishment of a fluid flux in the oviducts. Several aspects relating to the motion of flagella and cilia are still under debate, and computer simulations can be of help in clarifying these issues. An example of the use of finite element modeling to simulate axoneme bending has been published by Cibert and Heck [23]. These investigators studied the 'deviated bending' of axonemata in relation to the geometry of the axonemal doublets. In the mammalian inner ear, stereocilia are involved in the mechanoelectric transduction of acoustic waves (mechanical vibrations of the basilar membrane of Corti's organ) into neural signals. The cells bearing these stereocilia are referred to as inner-ear hair cells. Mechanically gated ion channels located near the apices of the stereocilia play a central role in the transduction process. Their opening probability is regulated by the deflection of the hair bundle (stereocilia). The magnitude of the deflection is a function of the geometrical and material properties of the bundle. Several measurements based on push or pull experiments have yielded information about the stiffness of stereocilia. But unfortunately, these experiments cannot correlate stiffness with other bundle parameters, such as the number of rows of stereocilia, gradations in the height of stereocila, or

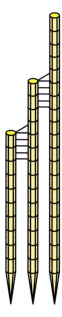

**Figure 12.4**  Basic unit of the inner-ear stereocilia and their interconnection as modeled by finite element method. The simulation was used to identify the contribution of individual bundle components to the overall stiffness.

inter-stereocilial spacing. A finite element analysis is well adapted to study the micromechanics of hair bundles since it can incorporate sophisticated geometries and complex material properties. Several such studies have been conducted among and others by Duncan and Grant and Silber et al. [24, 25]. In these investigations, not only single stereocilia but also complete and interconnected hair bundles were modeled (Figure 12.4).

The influence of various geometric, link, and material parameters on bundle stiffness and deformation shape were evaluated. A comprehensive review of the use of the finite element methodology in biology, which includes a detailed description of the simulation of inner-ear mechanics has been compiled by Kolston [26]. For a recent review of the mechanics of the inner ear, the reader is referred to the publication of Fettiplace and Hackney [27].

## 12.4.3 Cell nuclei

As distinct from bacteria, eukaryotic cells are characterized by the presence of a membrane-bound nucleus, which houses the DNA and is closely connected to the cytoskeleton. When the cell membrane is under stress, or when the cytoskeleton undergoes remodeling, the nucleus deforms. It has been suggested that this deformation of the nucleus could elicit a change in the packing of the DNA, which might influence gene regulation [28]. To better understand this deformation phenomenon, the mechanical properties of cell nuclei have been monitored either using micropipette-aspiration techniques or by applying micropipette-induced local strains to cells. In 2002, similar measurements were derived by compressing individual endothelial cells between glass microplates [29]. With the aid of these measurements, the investigators constructed a finite element model whose sensitivity to variations in material properties was explored. The finite element simulations also permitted the investigators to track the origin of the observed nonlinear behavior. By selecting a similar Young's modulus for the cytoplasm and the nucleus, they demonstrated the nonlinearity of the force-deformation curves to be attributable to the material behavior of the cell and not to the presence of a nucleus.

## 12.4.4 Micro-organisms

Since the mechanical properties of infectious agents such viruses and bacteria reflect the molecular composition of their envelopes, they probably contribute significantly to their infective potential. Nevertheless, these properties are poorly comprehended. Indeed, even if the stiffness of micro-organisms can be measured by relatively simple means, for example by atomic force microscopy, their complicated shapes render an interpretation of the data and the derivation of a Young's modulus almost impossible. However, if the geometry of the tip and the spring constant of the cantilever are known, the 3D shape of a micro-organism can be deduced from its image formed in the atomic force microscope, and this information can be used in conjunction with finite

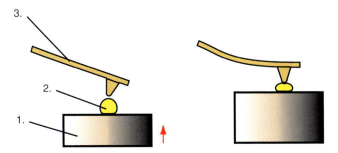

**Figure 12.5**  Measurement of the mechanical properties of a microscopic sample by atomic force microscopy. The piezo–electric crystal (1) which holds the sample (2) is extended until the atomic force microscope cantilever deforms (3) to a predefined level. Knowing the cantilever deformation and the piezo position, the indentation of the tip into the sample can be calculated.

element simulations to deduce the Young's modulus of its envelope (Figure 12.5).

For example, Kol et al. [30] initially measured the stiffness (in N/m) of murine leukemia virus particles by atomic force microscopy. They then modeled the shape of the virus using the finite element method and simulated the indentation process. The Young's modulus of the virus was then adjusted and the analysis repeated until the calculated rigidity fitted the measured one. By these means, differences in the mechanical properties of mature and immature murine leukemia virus particles were established. Using a similar approach (finite element modeling in conjunction with atomic force microscopy), Young's modulus has been determined for the cowpea chlorotic mottle viruses, $\phi$-phages, and the hyphae of *Aspergillus niduland* and *Saccaromyces cerevisiae*.

## 12.4.5 Single cells

An elegant use of the finite element analysis to account for the complex shape of certain unicellular organisms has been documented by Hamm et al. [31]. The reported study was motivated by the striking similarity between certain diatoms and statically sophisticated, man-made constructions that combined the properties of rigidity and lightness. Diatoms are single-celled algae bearing two hard siliceous coverings (frustules), which fit together in a box-and-lid-like fashion and probably serve as a protective physical barrier against mechanical stress. The investigators suspected that the shape of the coverings had been evolutionarily optimized to support the highest possible stress with the minimal amount of material. To test the validity of this postulated physical correlate, they first measured the forces that were required to break individual cells. They then used the finite element method to model the diatom frustules. By simulating the crushing procedure, they deduced the material properties of the frustules, the maximal stress that could be endured before breakage, and the stress profile as a function of mechanical loading. A similar experimental and analytical approach was adopted by Peeters et al. [32] to assess the mechanical and mechanical-failure properties of mammalian cells attached to a hard substrate. Under physiological conditions, living cells are continuously subjected to deforming mechanical forces, which influence several processes such as their growth, differentiation, survival, and gene expression. A knowledge of the mechanical properties of cells is thus essential to understand their biological activities. To asses the mechanical and mechanical-failure properties of their cells, Peeters et al. combined compression experiments with a finite element analysis to estimate Young's modulus. By comparing the experimental data with the simulated values, the material properties of the sample were deduced. Numerous other publications likewise describe how the finite element method can be used to explore the interaction between individual cells and a suitable measuring device, such as an atomic force microscope, a cytoindenter, or a magnetic twisting cytometer. A more sophisticated finite element model of living cells has been established by Herant et al. [33]. In their study, not only the mechanical properties but also dynamic behavior of Europhiles were simulated under the influence of various physical parameters. Their cell model, which was build up of an aqueous solvent phase (the hyaloplasm), a cytoskeletal network and the plasma membrane, permitted the investigators

**Figure 12.6** Schematic description of the phagocytic process as it has been simulated by Herant et al. The simulation permitted, among others, to monitor the elastic and viscous contributions to the surface tension.

to simulate the behavior of neutrophils upon aspiration into a micropipette, including their motion and the dynamics of pseudopod formation. Using their simulated data, the investigators were able to confirm or refute their different hypotheses. More recently, Herant et al. [34] have simulated the process of phagocytosis in neutrophils, likewise using a sophisticated finite element model (Figure 12.6).

Two other remarkable examples of the complexity of the models that can be established by a finite element analysis are furnished by the publications of Rubinstein and Jacobson and Bottino et al. [35, 36]. In these studies, motile cells were simulated using model parameters such as cell-to-substrate adhesivness, the transport of cytoskeletal proteins, the polymerization and depolymerization of cytoskeletal elements, and the contraction and protrusion processes of the cytoskeleton. Through these simulations, the observed shapes, forces, and movements of the living cells were reproduced. The means whereby biochemical and mechanical signals induce dynamic cellular processes, and the ways in which these transduction processes can be modeled by computer simulations, have been recently reviewed by Tracqui [37]. The finite element method has been used to simulate the mechanical behavior of numerous other cell types, including plant cells, leukocytes, osteoblasts, myoblasts, chondrocytes, and endothelial cells in conjunction with various modes of mechanical deformation, such as poking (plant cells), subjection to shear-flow conditions (endothelial cells and leukocytes), and indentation with the tip of an atomic force microscope (osteoblasts and endothelial cells).

## 12.4.6 Embryology and cell division

This final section deals with the finite-element modeling of several dynamic processes occurring during embryonic development, such as cell division and migration. The sequence of events that begins with a single fertilized ovum and culminates in a complex organism is still shrouded in mystery. The breaches in our knowledge span the fields of genetics, molecular biology, biochemistry, and also biomechanics. During the initial stage of development, the fertilized ovum, or zygote, transforms into a multicellular aggregate, the blastula, by cleavage, which is a variation of the mitotic process whereby all cells arise. Cleavage and cell division are putatively initiated by the formation of a contractile ring beneath the cell surface. In cooperation with molecular motors, this contractile ring of actin microfilaments causes the formation of a groove between the two nascent daughter cells and eventually pinches them off. This model originated during the last century and since then has been extended and refined [38]. However, the precise manner in which the cytoplasm divides during cell division is still a matter of debate, since different mechanisms of ingression can lead to similar results (Figure 12.7).

Several investigators have used a finite element analysis to test different hypotheses. The reliability of such analyses depends, of course, on a knowledge of the material properties of the cell in question and on the boundary conditions. Hence, the physical properties of cells and embryos have been subject to

**Figure 12.7** Model of the cell division with formation of an actin ring (red) and microtubule bundles (green). This simulation tested the different models explaining the mechanisms of cell division.

scrutiny. Various techniques have been employed for this purpose. These include (among others) micropipette aspiration, atomic force microscopy, laser-tracking microrheology, needle poking, and magnetic twisting cytometry. A highly sophisticated finite element simulation of the first cleavage division in the sea urchin embryo has been elaborated by He and Dembo [38]. Their model simulated the flow of the hyaloplasm, the polymerization and depolymerization of the cytoskeleton, the friction existing between the cytoskeletal network and the aqueous phase of the cytoplasm, and the viscosity as well as the contractility of the cytoskeletal network. It has been used to address several parameters such as the role of cytoskeletal flow during the formation of the contractile ring and the mechanics of the latter.

As the process of cleavage continues and the total number of cells increases, an almost spherical blastula is formed, which eventually invaginates to yield a gastrula (Figure 12.8).

Several mechanisms have been proposed to account for the shape changes occurring during gastrulation [39]. The study conducted by Davidson et al. illustrates particularly well the great potential of the finite element analysis in this field. Using optical sections of a *Lytechinus pictus* blastula, these investigators modeled its 3D shape. Since the material properties of the different components of the epithelial template were not completely known,

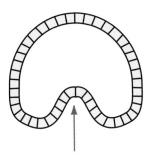

**Figure 12.8** Invagination process during gastrulation and neurulation. The finite element model probed the different hypothesis underlying the invagination.

they were set using values that had been determined by other investigators using various systems and methodologies. These included the compression of living cells between parallel plates, the suctioning of neutrophils and endothelial cells, and the monitoring of low-frequency dynamic modulii in purified cytoskeletal components. The 'completed' epithelial template was then used as a model to test five of the proposed mechanisms of invagination. By adjusting the mechanical properties of the different embryonic components, investigation could be achieved according to each of the tested mechanisms. These simulations revealed that each hypothesis should have a specific combination of mechanical properties. By determining the requisite material properties for other prospective mechanisms, their number could be reduced to the few that actually elicited invagination. For a comprehensive, biomechanically oriented review of gastrulation in different species, the reader is referred to Ref. [40].

In species with an axial nervous system, gastrulation is followed by neurulation, which involves the formation of the tubular rudiment of the nerve cord. Both gastrulation and neurulation involve a folding and reshaping of the epithelial sheets of the embryo (Figure 12.8). Although the outcomes of the tissue movements and rearrangements involved in gastrulation and neurulation differ, the underlying mechanisms are probably comparable and subject to similar constraints. For this reason, identical modeling techniques have been used to simulate each process. In 1993, Clausi and Brodland [41] simulated neurulation using a approach similar to that adopted for gastrulation. The neuronal plate, namely, the part of the embryonic epithelium that give rise to the neural tube by neurulation, consists of thousands of cells. Since hardware limitations would render impractical the individual modeling of all cells, the investigators simulated small patches of these with like volumes. Each simulated patch contained apical and proximal contractile structures to mimic the actin and tubulin filaments. The neural plate was simulated in both the dorsal and the transverse planes. The investigators performed several simulations, each of which was designed to test a particular hypothesis concerning the forces that drive neurulation and to gauge the susceptibility of the mechanical system to changes in properties of tissue.

Finite elements have also been used to study such embryological phenomena as cells sorting and checkerboard patterning. If two different types of embryonic cells are artificially dissociated, randomly mixed, and reaggregated, they spontaneously associate to re-establish a coherent homogeneous tissue. One explanation of this phenomenon is furnished by the differential adhesion hypothesis, which assumes cell movement to be driven by differences in cell–cell adhesion (reviewed in Ref. [42]). Alternatively, the phenomenon could be explained by the differential tension hypothesis. This tenet has been developed by the finite element modeling of mechanical interactions between cells within multicellular aggregates [43]. According to this model, microfilaments lying close to the cell surface can generate a contractile force that would influence cell–cell adhesion [44].

 ## 12.5 CONCLUSION

Owing to rapid developments in atomic force microscopy and related techniques, it is now becoming possible to measure the mechanical properties of biological material on the nano-to-micrometer scale with greater ease and less expense than heretofore. However, the data gleaned from these apparatuses are seldom open to a simple interpretation or amenable to the usual analytical tools, owing partially to the complex geometries of the samples and the testing devices, and to structural inhomogeneities in the former. These conditions give rise to marked nonlinearities in the mathematical model, which can be resolved only by heavy numerical manipulations. Amongst the tools available for solving these difficulties (such as the Monte Carlo, finite difference, or molecular dynamics techniques), the finite element method offers several advantages. It can model almost any geometry or material property and can simulate dynamic as well as static conditions. It can handle contacts and has multiphysics capabilities, namely, it can deal with interactions between mechanical, thermal, or electromagnetic processes. Furthermore, several affordable finite element packages are available in the market (http://homepage. usask.ca/~ijm451/finite/fe_resources/fe_resources.html), and their use does not usually require special programming or engineering skills. The latter in particular has perhaps contributed most substantially to the success of the method in the nonengineering scientific community.

## Bibliography

[1] Courant, R. (1923), On a convergence principle in calculus of variation (German), Berlin, Köln Gesellschaft der Wissenschaften zu Göttingen Wachrichten.

[2] Clough, R. W. (1960), The finite element in plan stress analysis, *Proc. the 2nd ASCE Conf. on Electron. Comput.*, Pittsburgh, PA.

[3] Melosh, R. J. (1963), Basis for derivation of matrices for the direct stiffness method, *AIAA J.*, 1, 1631–1637.

[4] Samuelson, A. and Zienkiewicz, O. C. (2006), History of the stiffness method, *Int. J. Numer. Methods Eng.*, 67, 149–157.

[5] Clough, R. W. (2004), Early history of the finite element method from the view point of a pioneer, *Int. J. Numer. Methods Eng.*, 60, 283–287.

[6] Thomee, V. (2001), From finite differences to finite elements – a short history of numerical analysis of partial differential equations, *J. Comput. Appl. Math.*, 128, 1–54.

[7] Moaveni, S. (1999), Finite Element Analysis, Prentice Hall. Upper Saddle River, New Jersey 07458.

[8] Gmur, T. (2000), Methode des elements finis en mecanique des structures, Presses Polytechniques et Universitaires Romandes.

[9] Adams, V. and Askenazi, A. (1999), Building Better Products with Finite Element Analysis, OnWord Press, Santa Fe, NM 87505-4835, USA.

[10] Van Vliet, K. J. and Bao, G. (2003), The biomechanics toolbox, experimental approaches for living cells and biomolecules, *Acta Mater.*, 511, 5881–5905.

[11] Bao, G. and Suresh, S. (2003), Cell and molecular mechanics of biological materials, *Nat. Mater.*, 211, 715–725.

[12] Boal, D. (2002), 'Mechanics of the Cell', Cambridge University Press, Cambridge, UK.

[13] Hansen, J. C., Skalak, R., Chien, S., and Hoger, A. (1997), Spectrin properties and the elasticity of the red blood cell membrane skeleton, *Biorheology*, 34, 327–348.

[14] De Matteis, M. A. and Morrow, J. S. (2000), Spectrin tethers and mesh in the biosynthetic pathway, *J. Cell Sci.*, 113, 2331–2343.

[15] Tang, Y., Cao, G., Chen. X., Yoo, J., Yethiraj, A., and Cui, Q. (2006), A finite element framework for studying the mechanical response of macromolecule, Application to the gating of the mechanosensitive channel MscL, *Biophys. J.*, 91, 1248–1263.

[16] Nogales, E., Whittaker, M., Milligan, R. A., and Downing, K. H. (1999), High-resolution model of the microtubule, *Cell*, 96, 79–88.

[17] Vinckier, A., Dumortier, C., Engelborghs, Y., and Hellemans, L. (1996), Dynamical and mechanical study of immobilized microtubules with atomic force microscopy, *J. Vac. Sci. Technol. B*, 14, 1427–1431.

[18] Kurachi, M., Hoshi, M., and Tashiro, H. (1995), Buckling of a single microtubule by optical trapping forces – direct measurement of microtubule rigidity, *Cell Motil. Cytoskeleton*, 30, 221–228.

[19] Kis, A., Kasas, S., Babic, B., Kulik, A. J., Benoit, W., Briggs, G. A. et al. (2002), Nanomechanics of microtubules, *Phys. Rev. Lett.*, 89, 248101, 1–4.

[20] Kasas, S., Cibert, C., Kis, A., De Los Rios, P., Riederer, B. M., Forro, L. et al. (2004), Oscillation modes of microtubules, *Biol. Cell*, 96, 697–700.

[21] Kasas, S., Kis, A., Riederer, B. M., Forro, L., Dietler, G., and Catsicas, S. (2004), Mechanical properties of microtubules explored using the finite elements method, *Chemphyschem*, 5, 252–257.

[22] Schaap, I. A., Carrasco, C., de Pablo, P. J., MacKintosh, F. C., and Schmidt, C. F. (2006), Elastic response, buckling, and instability of microtubules under radial indentation, *Biophys. J.*, 91, 1521–1531.

[23] Cibert, C. and Heck, J. V. (2004), Geometry drives the deviated-bending of the bi-tubular structures of the 9+2 – axoneme in the flagellum, *Cell Motil. Cytoskeleton*, 59, 153–168.

[24] Duncan, R. K. and Grant, J. W. (1997), A finite-element model of inner ear hair bundle micromechanics, *Hear. Res.*, 104, 15–26.

[25] Silber, J., Cotton, J., Nam, J. H., Peterson, E. H., and Grant, W. (2004), Computational models of hair cell bundle mechanics, III. 3D utricular bundles, *Hear. Res.*, 197, 112–130.

[26] Kolston, P. J. (2000), Finite-element modelling, a new tool for the biologist, *Philos. Trans. R. Soc. London Ser. A – Math. Phys. Eng. Sci.*, 358, 611–631.

[27] Fettiplace, R. and Hackney, C. M. (2006), The sensory and motor roles of auditory hair cells, *Nat. Rev. Neurosci.*, 7, 19–29.

[28] Gimbrone, M. A., Jr, Resnick, N., Nagel, T., Khachigian, L. M., Collins, T., and Topper, J. N. (1997), Hemodynamics, endothelial gene expression, and atherogenesis. *Ann. N. Y. Acad. Sci.*, 811, 1–10.

[29] Caille, N., Thoumine, O., Tardy, Y., and Meister, J. J. (2002), Contribution of the nucleus to the mechanical properties of endothelial cells, *J. Biomech.*, 35, 177–187.

[30] Kol, N., Gladnikoff, M., Barlam, D., Shneck, R. Z., Rein, A., and Rousso, I. (2006), Mechanical properties of murine leukemia virus particle: effect of maturation, *Biophys. J.*, 91, 767–774.

[31] Hamm, C. E., Merkel, R., Springer, O., Jurkojc, P., Maier, C., Prechtel, K. et al. (2003), Architecture and material properties of diatom shells provide effective mechanical protection, *Nature*, 421, 841–843.

[32] Peeters, E. A., Oomens, C. W., Bouten, C. V., Bader, D. L., and Baaijens, F. P. (2005), Mechanical and failure properties of single attached cells under compression, *J. Biomech.*, 38, 1685–1693.

[33] Herant, M., Marganski, W. A., and Dembo, M. (2003), The mechanics of neutrophil: synthetic modeling of three experiments, *Biophys. J.*, 84, 3389–3413.

[34] Herant, M., Heinrich, V., and Dembo, M. (2006), Mechanics of neutrophil phagocytosis, experiments and quantitative models, *J. Cell Sci.*, 119, 1903–1913.

[35] Rubinstein, B. and Jacobson, K. (2005), Multiscale two-dimensional modeling of a motile simple-shaped cell, *Multiscale Model. Simul.*, 3, 413–439.

[36] Bottino, D., Mogilner, A., Roberts, T., Stewart, M., and Oster, G. (2002), How nematode sperm crawl, *J. Cell Sci.*, 115, 367–384.

[37] Tracqui, P. (2006), Mechanical instabilities as a central issue for insilico analysis of cell dynamics, *Proc. IEEE*, 94, 710–724.

[38] He, X. Y. and Dembo, M. (1997), On the mechanics of the first cleavage division of the sea urchin egg, *Exp. Cell Res.*, 233, 252–273.

[39] Davidson, L. A., Koehl, M. A., Keller, R., and Oster, G. F. (1995), How do sea-urchins invaginate – using biomechanics to distinguish between mechanisms of primary invagination, *Development*, 121, 2005–2018.

[40] Keller, R., Davidson, L. A., and Shook, D. R. (2003), How we are shaped: the biomechanics of gastrulation, *Differentiation*, 71, 171–205.

[41] Clausi, D. A. and Brodland, G. W. (1993), Mechanical evaluation of theories of neurulation using computer-simulations, *Development*, 118, 1013–1023.

[42] Foty, R. A. and Steinberg, M. S. (2004), Cadherin-mediated cell–cell adhesion and tissue segregation in relation to malignancy, *Int. J. Dev. Biol.*, 48, 397–409.

[43] Brodland, G. W. and Chen, H. H. (2000), The mechanics of heterotypic cell aggregate: insights from computer simulations, *J. Biomech. Eng. – Trans. ASME*, 122, 402–407.

[44] Brodland, G. W. (2002), The differential interfacial tension hypothesis (DITH): a comprehensive theory for the self-rearrangement of embryonic cells and tissues, *J. Biomech. Eng. – Trans. ASME*, 124, 188–197.

# BEAM BENDING

## Contents

## > A.1.1 BEAM BENDING

### A.1.1.1 Supported beam at two ends

The force transducer of AFM is a cantilever sensor and the degree of its bending angle is related to its bending deflection. First, for those who are not familiar with beam bending mechanics, an introduction to beam-bending mechanics is given below.

Suppose a rectangular beam of length $L$, width $w$, and thickness $t$ is placed on two supports at the left and right ends, respectively, A and B with an application of a point load of $P$ at $x = a$ from A, as shown in Figure A.1.1.

A point load is an idealized case of an application of force (or equivalently weight) within a small area. If the force is applied over a wide area, it is treated as a distributed force with an intensity of $q(x)$ at position $x$. We like to know how to estimate the deflection of the beam at a distance $x$ from the left cnd. In mechanics,

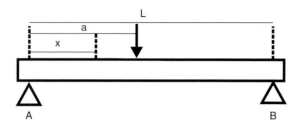

**Figure A.1.1**   A beam is supported at its two ends at A and B and a weight is applied at a distance $a$ from A.

these types of problems are treated routinely by considering the shear force and the moment at each respective points. The support at A and B could be one of the three types, namely, roller support, where only the vertical reaction $(R_A)$ has a finite value; pin support, where the vertical and horizontal reactions $(H_A)$ have finite values; and fixed support, where in addition to the vertical and horizontal reactions, the moment $(M_A)$ is also finite. For example, in the point load case given above,

*Roller support*: The beam can move horizontally on the support. Vertical stress at A is finite, but both the horizontal stress and moment are zero.

$$R_A = -F\frac{L-a}{L}, \ R_B = -F\frac{a}{L} \qquad \text{(A.1.1)}$$

*Pin support*: The beam cannot move horizontally. Both vertical and horizontal stresses are finite, but moment at A is zero.

$$R_A = -F\frac{L-a}{L}, \ H_A = -F\sin\theta \qquad \text{(A.1.2)}$$

where $\theta$ is the angle of deflection from the horizontal line of the beam at A.

*Fixed support*: The beam has no freedom of either horizontal or rotational move at A. Vertical and horizontal stresses and the moment are nonzero at A.

$$R_A = -F\frac{L-a}{L}, \ H_A = -F\sin\theta \qquad \text{(A.1.3)}$$

$$M_A = Fa, \ M_B = F(L-a) \qquad \text{(A.1.4)}$$

where $\theta$ is the angle of deflection from the horizontal line of the beam at A.

The relation between the moment and the curvature of the beam at the position of $x$ is obtained from the equilibrium consideration as explained in Figure A.1.2.

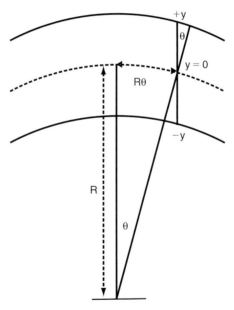

**Figure A.1.2**  2D geometrical diagram of beam bending. A straight beam (length $= L$) of a rectangular cross-section (width $= w$ and thickness $= t$) is bent by the application of a weight $P$ at a position $x = 0$. The beam is assumed to be bent in a circular shape with the radius of curvature $R$. The dashed line running through the center of the beam is called the neutral line, and the beam is compressed above and extended below it. Thus along the neutral line, the beam has no deformation. We consider the compression and extension of the beam at a position $x$, with an angle $\theta$ defining the length of the arc from $x = 0$ to $x$.

When the radius of beam curvature is $R$ and the angle subtended by the contour length of a sample piece of the beam is $\theta$, the original length of the beam piece is $R\theta$, and the elongation in the beam above and below the neutral line are, respectively, $y\theta$ and $-y\theta$, which are, in terms of strain, $y/R$ and $-y/R$. The stress $\sigma$ is, therefore, $yY/R$ and $-yY/R$, where $Y$ is Young's modulus of the beam material. The stress, therefore, is canceled between the upper and lower halves of the beam. Although the beam is free from stress due to bending, the moment at $x$ remains finite because it is $y \times yY/R$ and $(-y) \times (-yY/R)$, respectively, in the upper and the lower halves of the beam. By integrating the resultant moment of $y^2Y/R$ over the cross-sectional area of the beam, we obtain,

$$M = \frac{Y}{R} \int\limits_{-w/2}^{w/2} \int\limits_{-t/2}^{t/2} y^2 \, \mathrm{d}y \mathrm{d}z = \frac{YI}{R} \qquad (A.1.5)$$

where $w$ and $t$ are the width and the thickness of the beam, and

$$I = \int\limits_{-w/2}^{w/2} \int\limits_{-t/2}^{t/2} y^2 \mathrm{d}y \mathrm{d}z \qquad (A.1.6)$$

is called the cross-sectional moment of the second order.

Since the radius of curvature is reciprocal of the curvature, which is given by

$$\frac{1}{R} = \frac{\dfrac{\mathrm{d}^2 y}{\mathrm{d}x^2}}{\left\{ 1 + \left( \dfrac{\mathrm{d}y}{\mathrm{d}x} \right)^2 \right\}^{3/2}} \qquad (A.1.7)$$

which is reduced to $\mathrm{d}^2 y/\mathrm{d}x^2$ if the beam is always assumed to be placed horizontally in the beginning.

The deflection is calculated by solving the following differential equation.

$$\frac{\mathrm{d}^2 y}{\mathrm{d}x^2} = \frac{M}{YI} \tag{A.1.8}$$

Next, we obtain the shear force acting on the cross-section of the beam at $x$ and the moment there. First, the shear force is the sum of the action and reaction, which equals to zero. Second, the moment at $x$ is the difference between the moment exerted by $R_A$ at the position $x$ and that exerted by $F$ at $a$ at the distance of $-(x - a)$. First,

$$R_A = \frac{F(L - a)}{L} \tag{A.1.9}$$

$$R_B = -\frac{Fa}{L} \tag{A.1.10}$$

Because the definition of the moment is (force) × (distance), for a small displacement $\mathrm{d}x$, the moment is $\mathrm{d}M = M\mathrm{d}x$. Then, the basic relation between $F$ and $M$ is,

$$\frac{\mathrm{d}M}{\mathrm{d}x} = F(x), \text{ therefore, } M = \int_0^x F(x)\mathrm{d}x \tag{A.1.11}$$

By applying these relations, the results for the case in Figure A.1.2 are given as follows

$$F_S = \frac{F(L - a)}{L}, \text{ (for } 0 \le x \le a) \tag{A.1.12}$$

$$F_S = \frac{F(L - a)}{L} - F, \text{ (for } a \le x \le L) \tag{A.1.13}$$

$$M = \frac{F(L - a)}{L}x, \text{ (for } 0 \le x \le a) \tag{A.1.14}$$

$$M = \frac{F(L - a)}{L}x - F(x - a) \text{ (for } a \le x \le L) \tag{A.1.15}$$

The formulae for bending moment are different in two regions, we have,

$$YI\frac{d^2y}{dx^2} = \frac{F(L-a)}{L}x, \text{ (for } 0 \le x \le a) \qquad \text{(A.1.16)}$$

$$YI\frac{d^2y}{dx^2} = \frac{F(L-a)}{L}x - F(x-a), \text{ (for } a \le x \le L)$$
$$\text{(A.1.17)}$$

By integrating we obtain,

$$YI\frac{dy}{dx} = \frac{F(L-a)}{2L}x^2 + C_1, \text{ (for } 0 \le x \le a) \qquad \text{(A.1.18)}$$

$$YI\frac{dy}{dx} = \frac{F(L-a)}{2L}x^2 - \frac{F(x-a)^2}{2} + C_2, \text{ (for } 0 \le x \le a)$$
$$\text{(A.1.19)}$$

Since the two parts of the beam are smoothly connected at $x = a$, $C_1 = C_2 = C$.

$$YIy = \frac{F(L-a)}{6L}x^3 + Cx + C_3, \text{ (for } 0 \le x \le a) \qquad \text{(A.1.20)}$$

$$YIy = \frac{F(L-a)}{6L}x^3 - \frac{F(x-a)^3}{6} + Cx$$
$$+ C_4, \text{ (for } a \le x \le L) \qquad \text{(A.1.21)}$$

The deflection is same at $x = a$, therefore, $C_3 = C_4 = C'$, and at $x = 0$, $y = 0$ and at $x = L$, $y = 0$, therefore,

$$C = -\frac{F(L-a)[L^2 - (L-a)^2]}{6L}, \text{ and } C' = 0 \qquad \text{(A.1.22)}$$

Finally, we obtain the magnitude of vertical deflection of the beam as,

$$y = \frac{F(L-a)x}{6LYI}(L^2 - (L-a)^2 - x^2), \text{ (for } 0 \le x \le a) \quad (A.1.23)$$

$$y = \frac{F(L-a)x}{6LYI}(L^2 - (L-a)^2 - x^2) + \frac{F(x-a)^3}{6YI},$$

$$(\text{for } a \le x \le L) \quad (A.1.24)$$

and the slope of the beam bending as follows.

$$\frac{dy}{dx} = \frac{F(L-a)}{6LYI}(L^2 - (L-a)^2 - 3x^2), \text{ (for } 0 \le x \le a) \quad (A.1.25)$$

$$\frac{dy}{dx} = \frac{F(L-a)}{6LYI}(L^2 - (L-a)^2 - 3x^2) + \frac{F(x-a)^2}{2YI},$$

$$(\text{for } a \le x \le L) \quad (A.1.26)$$

## A.1.1.2 Cantilever bending

A cantilever has a fixed support at one end and a free end at the other. Suppose we have a cantilever of length $L$, width $w$, and thickness $t$ on which a vertical point force of $F$ is acting at a position $x$ from the free end. No reaction or moment is acting on the free end. On the fixed end, we have,

$$R_A = F_S, \text{ and } M_A = F \times (L - x) \quad (A.1.27)$$

and at point $x$, the shear force cancels out, but

$$M = F \times x \quad (A.1.28)$$

Consequently, we have the following differential equation, with the boundary conditions given explicitly.

$$\frac{d^2y}{dx^2} = \left(\frac{1}{YI}\right)Fx \quad (A.1.29)$$

with the boundary condition such as, $y(L) = 0$ and $(dy/dx)_{x=L} = 0$.

By integrating the above equation from $x = 0$ to $x$, we obtain,

$$y = \frac{FL^3}{6YI}\left[2 - 3\left(\frac{x}{L}\right) + \left(\frac{x}{L}\right)^3\right] \qquad (A.1.30)$$

$$\frac{dy}{dx} = \frac{FL^3}{6YI}\left[-\left(\frac{3}{L}\right) + \left(\frac{3x^2}{L^3}\right)\right] \qquad (A.1.31)$$

The deflection and the slope at the free end ($d$ and $(dy/dx)_{x=0}$) are,

$$(y)_{x=0} = d = \frac{FL^3}{3YI} \qquad (A.1.32)$$

$$\left(\frac{dy}{dx}\right)_{x=0} = \frac{-FL^2}{2YI} \qquad (A.1.33)$$

## A.1.1.3 Distributed force

Figure A.1.3 shows the weight distributed over the beam with the intensity of $q(x)$ at position $x$.

In this case, the shear force on the right-hand segment at the cross-section at $x$ is $F = qx$, downward shear force being taken as a positive force. The shearing effect of all the point force $q(x)$ is accumulated at the cross-section at $x$. The subsequent calculation to obtain a formula for beam deflection is the same as described above, starting from the following equation with appropriate boundary conditions.

$$\frac{d^2y}{dx^2} = \frac{M}{YI} = \frac{1}{YI}Fx = \frac{1}{YI}qx^2 \qquad (A.1.34)$$

The result is,

$$y = \frac{qL^4}{24YI}\left[\frac{x}{L} - 2\left(\frac{x}{L}\right)^3 + \left(\frac{x}{L}\right)^4\right] \qquad (A.1.35)$$

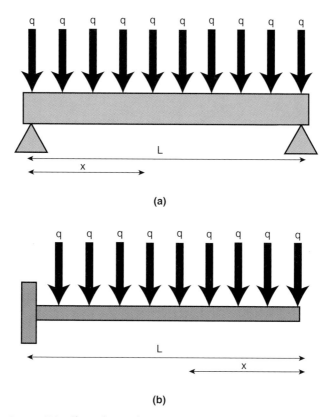

**Figure A.1.3** Distributed weight $q$ on a beam (a) and on a cantilever (b) with definitions of $L$ and $x$.

$$\frac{\mathrm{d}y}{\mathrm{d}x} = \frac{qL^3}{24YI}\left[1 - 6\left(\frac{x}{L}\right)^2 + 4\left(\frac{x}{L}\right)^3\right] \qquad (A.1.36)$$

and the maximum values are,

$$y_{\max} = \frac{5}{384}\frac{qL^4}{YI} \qquad (A.1.37)$$

$$\left(\frac{\mathrm{d}y}{\mathrm{d}x}\right)_{\max} = \frac{qL^3}{24YI} \qquad (A.1.38)$$

For a cantilever with the same geometry as above,

$$y = \frac{qL^4}{24YI}\left[3 - 4\left(\frac{x}{L}\right) + \left(\frac{x}{L}\right)^4\right] \qquad \text{(A.1.39)}$$

$$\frac{dy}{dx} = \frac{qL^3}{6YI}\left[\left(\frac{x}{L}\right)^3 - 1\right] \qquad \text{(A.1.40)}$$

$$y_{max} = \frac{qL^4}{8YI} \qquad \text{(A.1.41)}$$

$$\left(\frac{dy}{dx}\right)_{max} = -\frac{qL^3}{6YI} \qquad \text{(A.1.42)}$$

### A.1.1.4 Radius of curvature

The radius of curvature can be derived as follows. First, the relationship between an infinitesmal arc, $ds$, and the radius of curvature, $R$, is given in Figure A.1.4.

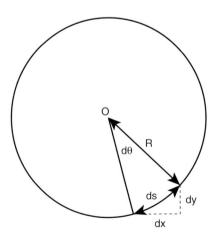

**Figure A.1.4**  Definition of $ds$ as the product of $R$ and $d\theta$.

The radius of curvature $R$ and the length of a small arc $ds$ are given as below (see Figure A.1.4 for reference).

Since $R d\theta = ds$,

$$R = \frac{ds}{d\theta} = (ds/dx)/(d\theta/dx) \tag{A.1.43}$$

We will obtain $ds/dx$ and $d\theta/ds$ to calculate $R$. First from $ds = \sqrt{dx^2 + dy^2}$,

$$\frac{ds}{dx} = \sqrt{1 + \left(\frac{dy}{dx}\right)^2} \tag{A.1.44}$$

Next, by definition of the derivative, $\tan\theta = dy/dx$, therefore by differentiating both sides of the equation with respect to $x$,

$$\frac{d\tan\theta}{d\theta}\frac{d\theta}{dx} = \frac{d^2y}{dx^2} \tag{A.1.45}$$

$$\frac{1}{\cos^2\theta}\frac{d\theta}{dx} = \frac{d^2y}{dx^2} \tag{A.1.46}$$

$$\frac{d\theta}{dx} = \cos^2\theta\frac{d^2y}{dx^2} = \frac{1}{1+\tan^2\theta}\frac{d^2y}{dx^2} = \frac{1}{1+\left(\frac{dy}{dx}\right)^2}\frac{d^2y}{dx^2} \tag{A.1.47}$$

$$R = \frac{ds}{d\theta} = \frac{\left[1 + \left(\frac{dy}{dx}\right)^2\right]^{3/2}}{\frac{d^2y}{dx^2}} \tag{A.1.48}$$

curvature $\kappa$ is the reciprocal of $R$.

$$\kappa = \frac{1}{R} \tag{A.1.49}$$

**A.1.2 Buckling**

We observe buckling phenomena in daily life when we compress a thin and long plastic plate axially. When the force is small, the plate remains straight as long as the compression force is small; however, it suddenly bent sideways to relax the stress. Or, while on board of an airplane, you open a water bottle and half empty it by drinking some water and recap it tightly and forget about it. Then after some hours, as the airplane starts descending, you will notice that parts of the bottle start popping in here and there. This is an example of 3D buckling. Buckling occurs suddenly and you may wonder how to formulate the constitutive equation for buckling.

Now, let us suppose that a beam of length $L$ having the bottom end free to pivot receives an axial compressive force from the top end. It has been known that if the force exceeds the Euler force, the beam buckles into a sinusoidal shape as shown in Figure A.1.5.

$$F_c = \pi^2 \frac{YI}{L^2} \tag{A.1.50}$$

Since once the beam starts buckling, little additional force is needed to further increase the bend, and the beam almost instantly collapses.

The expression for Euler force can be obtained in the following way by considering a beam under the axial force of $-F$ for case 1 in Figure A.1.5. The bending moment $M$ of the beam at distance $x$ from the fixed end of the beam when its lateral displacement at the other end is $\delta$ is,

$$M = -F(\delta - y) \tag{A.1.51}$$

The equation of beam bending is,

$$\frac{d^2 y}{dx^2} = -\frac{M}{YI} = \frac{F}{YI}(\delta - y) \tag{A.1.52}$$

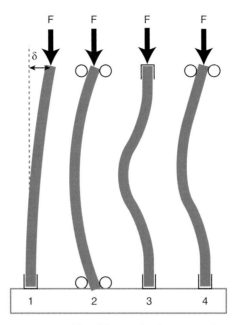

**Figure A.1.5** Two types of buckling of a beam under axial compression: (1) One end fixed and the other end free; (2) supported by roller supports on both ends; (3) both ends supported by fixed support; (4) one end fixed and the other end roller support.

The general solution of the above differential equation is given by the sum of trigonometric functions as below.

$$y = C_1 \sin \alpha x + C_2 \cos \alpha x + \delta, \ \ \alpha = \sqrt{\frac{F}{YI}} \qquad \text{(A.1.53)}$$

By applying a boundary condition at $x = 0$, $y = 0$, and $dy/dx = 0$, we determine two coefficients as below.

$$C_1 = 0, \ C_2 = -\delta \qquad \text{(A.1.54)}$$

Thus, the solution has the following simple form.

$$y = \delta(1 - \cos \alpha x) \qquad \text{(A.1.55)}$$

Further, since at $x = L$, $y = \delta$,

$$\delta = \delta(1 - \cos\alpha L), \quad \delta\cos\alpha L = 0 \qquad \text{(A.1.56)}$$

Excepting the trivial case of $\delta = 0$ where there is no deflection,

$$\cos\alpha L = 0, \text{ therefore, } \alpha L = (2n + 1)\frac{\pi}{2} \ (n = 0, 1, 2, \cdots)$$
$$\text{(A.1.57)}$$

Substituting the above result in $\alpha = \sqrt{F/YI}$, we obtain,

$$F = (2n + 1)^2 \frac{\pi^2 YI}{4L^2} \ (n = 0, 1, 2, \cdots) \qquad \text{(A.1.58)}$$

Since buckling takes place at the lowest value of F,

$$F = \frac{\pi^2 YI}{4L^2} \ (\text{Euler force}) \qquad \text{(A.1.59)}$$

For different types of boundary conditions, the following equation is obtained with different values of the fixity coefficient $k$ for each case of boundary conditions.

$$F_c = k\frac{\pi^2 YI}{L^2} \qquad \text{(A.1.60)}$$

$$\sigma_c = k\frac{\pi^2 Y}{\lambda^2} \qquad \text{(A.1.61)}$$

One end fixed, the other end free, $k = \dfrac{1}{4}$ \quad (A.1.62)

Both ends free to rotate, $k = 1$ \qquad (A.1.63)

Both ends fixed, $k = 4$ \qquad (A.1.64)

One end fixed, the other rotate, $k = 2.046$ \quad (A.1.65)

## A.1.3 BASICS OF LINEAR MECHANICS
## ACCORDING TO LANDAU AND LIFSHITZ

The textbook by Landau and Lifshitz [1] gives a concise and clear exposition of linear mechanics, which will be useful for those who are concerned with the applicable extent of linear mechanics. In mechanics, a general deformation of a solid body is expressed by the displacement vector, $u_i$, from the original position vector, $r_i$, with components, $x_i$, $(i = 1, 2, 3)$ to a new position $r'_i$ with components, $x'_i$. (In the following explanation, general summation rule of Einstein is used.)

$$u_i = x'_i - x_i \qquad (A.1.66)$$

If $u_i$ is given as a function of $x_i$ for all $i$, the deformation of the body is all solved. If the two points are very close to each other and the distance joining these two points is initially $dx_i$, and $dx'_i$ after deformation, $dx'_i = dx_i + du_i$. The distance between the two points before the deformation is $dL^2$ and that after the deformation is $dL'^2$.

$$dx'_i = dx_i + du_i \qquad (A.1.67)$$

$$dL^2 = dx_1^2 + dx_2^2 + dx_3^2 \quad dL^2 = dx_i^2 \qquad (A.1.68)$$

$$dL'^2 = dx_1'^2 + dx_2'^2 + dx_3'^2 \quad dL'^2 = dx_i'^2 = (dx_i^2 + du_i^2) \qquad (A.1.69)$$

Since $du_i = (\partial u_i / \partial x_k) dx_k$

$$dL'^2 = dL^2 + 2\frac{\partial u_i}{\partial x_k} dx_i dx_k + \frac{\partial u_i}{\partial x_k}\frac{\partial u_i}{\partial x_l} dx_k dx_l \qquad (A.1.70)$$

Since the summation is taken over both suffixes $i$ and $k$ in the second term on the right, this term can be put in the explicitly symmetrical form.

$$\left(\frac{\partial u_i}{\partial x_k} + \frac{\partial u_k}{\partial x_l}\right) dx_i dx_k \qquad (A.1.71)$$

In the third force, we interchange the suffixes $i$ and $l$. Then, $\mathrm{d}L'^2$ takes the final form.

$$\mathrm{d}L'^2 = \mathrm{d}L^2 + 2u_{ik}\mathrm{d}x_i\mathrm{d}x_k \qquad (A.1.72)$$

where the tensor $u_{ik}$ is defined as,

$$u_{ik} = \frac{1}{2}\left(\frac{\partial u_i}{\partial x_k} + \frac{\partial u_k}{\partial x_l} + \frac{\partial u_l}{\partial x_i}\frac{\partial u_l}{\partial x_k}\right) \quad u_{ik} = u_{ki} \qquad (A.1.73)$$

The tensor $u_{ik}$ is called the strain tensor, which is a symmetrical tensor according to its definition.

$$u_{ik} = u_{ki} \qquad (A.1.74)$$

Like any symmetrical tensor, $u_{ik}$ can be diagonalized at any given point. This means that, at any given point, we can choose coordinate axes (the principal axes of tensor) in such a way that only the diagonal components, $u_{11}, u_{22}, u_{33}$ of the tensor are nonzero. These are the principal values of the tensor and are denoted as $u^1, u^2$, and $u^3$. If the strain tensor is diagonalized, the length $\mathrm{d}L'^2$ becomes,

$$\mathrm{d}L'^2 = (\delta_{ik} + 2u_{ik})\mathrm{d}x_i\mathrm{d}x_k \qquad (A.1.75)$$
$$= (1 + 2u^1)\mathrm{d}x_1^2 + (1 + 2u^2)\mathrm{d}x_2^2 + (1 + 2u^3)\mathrm{d}x_3^2 \qquad (A.1.76)$$

If the strain is in $x_1$ direction only,

$$\mathrm{d}x_1'^2 = (1 + 2u^1)\mathrm{d}x_1^2 \qquad (A.1.77)$$

$$\mathrm{d}x_1' = \sqrt{(1 + 2u^1)}\,\mathrm{d}x_1 = (1 + u^1)\mathrm{d}x_1 \qquad (A.1.78)$$

The quantity $u^1$ is consequently equal to the relative extension $(\mathrm{d}x_i' - \mathrm{d}x_i)/\mathrm{d}x_i$ along the $i$th principal axis.

## Bibliography

[1] Landau, L. D. and Lifshitz, E. M. (1986), 'Theory of Elasticity', 3rd English ed. Butterworth-Heinemann.

**APPENDIX TWO**

# V-SHAPED CANTILEVER

## Contents

## A.2.1 V-SHAPED CANTILEVER

The theoretical treatment of V-shaped cantilever with respect to its spring constant has been treated by several groups, in which case the original proposition of the parallel beam approximation (PBA) by Albrecht et al. [1] and a theoretical confirmation of the PBA later by Sader are introduced [2]. First, the V-shaped cantilever was fabricated by Albrecht et al. [1] and in the same paper the spring constant of V-shaped cantilever was assumed to be equivalent to the rectangular cantilever having the same length and thickness but twice the width of the one arm of the original V-shaped one. Sader showed that this original approximation was correct within an allowable error if the length and the width are taken as shown in Figure A.2.1. In practice, the spring constant of the cantilever is determined by recording the thermal noise of the cantilever and by analyzing it according to the method proposed by Hutter and Beckhoffer [3] or by obtaining a force curve on pushing a calibrated cantilever [4].

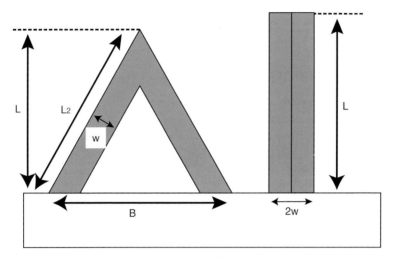

**Figure A.2.1** A schematic view of a V-shaped cantilever and a mechanically equivalent rectangular one.

## Bibliography

[1] Albrecht, T. R., Akamine, S., Carver, T. E., and Quate, C. F. (1990), Microfabrication of cantilever styli for the atomic force microscope, *J. Vac. Sci. Technol.*, A 8, 3386–3396.

[2] Sader, J. E. (1995), Parallel beam approximation for V-shaped atomic force microscope cantilevers. *Rev. Sci. Instrum.*, 66, 4583–4587.

[3] Hutter, J. L. and Bechhoefer, J. (1993), Calibration of atomic-force microscope tips, *Rev. Sci. Instrum.*, 64, 1868–1873.

[4] Torii, A., Sasaki, M., Hane, K., and Okuma, S. (1996), A method for determining the spring constant of cantilevers for atomic force microscopy, *Meas. Sci. and Technol.*, 7, 179–184.

**APPENDIX THREE**

# PERSISTENCE LENGTH AND KUHN STATISTICAL SEGMENT

---

## Content

---

## A.3.1   PERSISTENCE LENGTH AND KUHN STATISTICAL SEGMENT

For a polymer of length $L$, we will parameterize the path of the polymer along the chain as $s$, and define $t(s)$ to be the unit tangent vector to the chain at $s$, and $r$ to be the position vector along the chain.

$$t(s) = \frac{\partial r}{\partial s} \qquad (A.3.1)$$

and the end-to-end distance $R$ is,

$$R = \int_0^L t(s)\mathrm{d}s \qquad (A.3.2)$$

It can be shown that the orientation correlation function for a worm-like chain follows an exponential decay:

$$< t(s) \cdot t(0) > = < \cos\theta(s) > = e^{-s/p} \qquad (A.3.3)$$

where $p$ is by definition the polymer's characteristic persistence length. A useful value is the mean square end-to-end distance of the polymer:

$$< R^2 > = < \mathbf{R} \cdot \mathbf{R} > = < \int_0^L \mathbf{t}(s)\mathrm{d}s \times \int_0^L \mathbf{t}(s')\mathrm{d}s' > \quad \text{(A.3.4)}$$

$$= < \int_0^L \mathrm{d}s \times \int_0^L \mathbf{t}(s)\mathbf{t}(s')\mathrm{d}s' > \quad \text{(A.3.5)}$$

$$= < \int_0^L \mathrm{d}s \times \int_0^L \exp\left[-|s-s'|/p\right]\mathrm{d}s' > \quad \text{(A.3.6)}$$

$$= 2pL[1 - \frac{p}{L}(1 - e^{\frac{p}{L}})] \int_0^L \exp\left[-|s-s'|/p\right]\mathrm{d}s']$$

$$\text{(A.3.7)}$$

Note that in the limit of $L \gg p$, then $< R^2 > = 2pL$. Since $< R^2 > = np^2$ and $L = nL_K$, $np^2 = 2nL_Kp$, it follows that

$$p = 2L_K \quad \text{(A.3.8)}$$

# HERTZ MODEL

## Contents

## A.4.1 HERTZ MODEL

### A.4.1.1 Concentrated load

First, we will look at the result for the case of concentrated load at a single point on a flat, semi-infinitive surface. The function describing $z$-deformation at a radial distance $r$ as the result of point loading at $r = 0$ is obtained from the analysis based on the Boussinesq potential functions as described in Ref. [1], and the result is,

$$\bar{u}_z = \frac{1 - \nu}{2\pi G} \frac{P}{r} \qquad (A.4.1)$$

This function is then used to obtain an equation describing the degree of deflection due to an application of distributed load over a finite area on a flat surface.

### A.4.1.2 Distributed load

For a load applied on a nonzero finite area, the $z$-deflection is obtained by integrating the point-load function over the area with a varying degree of load intensity, $p(r, \theta)$. The cumulative effect of distributed load on the position designated as $B$ in Figure A.4.1 can be obtained by switching the variable from $r$ and $\theta$ to $s$ and $\phi$.

By replacing $r$ in the above equation and substituting $sd\phi ds$ as the surface element of integration and replacing $p(r, \theta)$ with $p(s, \phi)$, where $s$ and $\phi$ are defined in Figure A.4.1

$$\bar{u}_z = \frac{1 - \nu}{2\pi G} \int_S \int \frac{p(s, \phi)}{s} sd\phi ds = \frac{1 - \nu^2}{\pi Y} \int_S \int p(s, \phi) d\phi ds \tag{A.4.2}$$

For distributed loading, the following axi-symmetric formula with $n$ as a parameter to fit for different experimental situations gives solutions in closed form, where $a$, $p_0$, and $r$ are, respectively, the radius of the circle, distance from the center, and the load at the center of the circle, where $r = 0$.

$$p = p_0(1 - r^2/a^2)^n \tag{A.4.3}$$

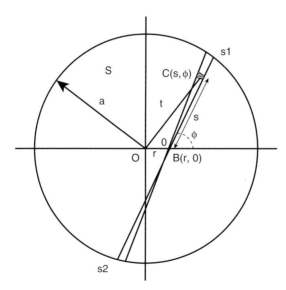

**Figure A.4.1** Scheme of integration of the point-load function over a distributed load of radius $a$ and with a circular symmetry. Reproduced from Figure 3.5 in Ref. [1] with permission.

The value of $n$ is given for two typical cases.

- Uniform pressure: $n = 0$
- Hertz pressure: $n = \frac{1}{2}$

**Hertz pressure (n $= \frac{1}{2}$)**

The cumulative effect of the pressure distribution on a particular point $B$ is given below and, the point-load equation is integrated over the entire circular region.

$$p(r) = p_0(a^2 - r^2)^{1/2}/a \tag{A.4.4}$$

To obtain the cumulative effect of all the pressure on B, the pressure at a distance $t$ from the center will be integrated with respect to $s$ and $\phi$. To do so, $(a^2 - r^2)$ in the pressure equation is replaced with $(a^2 - t^2)$, which is equal to $(a^2 - r^2) - (s^2 + 2rs\cos\phi)$, because $t^2 = r^2 + s^2 - 2rs\cos(\pi - \phi) = r^2 + s^2 + 2rs\cos\phi$.

By substituting,

$$\alpha^2 = a^2 - r^2 \tag{A.4.5}$$

$$\beta = r\cos\phi \tag{A.4.6}$$

we obtain,

$$a^2 - t^2 = \alpha^2 - 2\beta s - s^2 \tag{A.4.7}$$

By replacing $p(r)$ with $p(s,\phi)$, noting that $s_1$ is the positive root of,

$$\alpha^2 - 2\beta s - s^2 = 0 \tag{A.4.8}$$

Integration proceeds as follows.

$$\bar{u}_z(r) = \frac{1 - \nu^2}{\pi Y} \frac{p_0}{a} \int_0^{2\pi} d\phi \int_0^{s_1} (\alpha^2 - 2\beta s - s^2)^{1/2} ds \tag{A.4.9}$$

$$\int_0^{s_1} (\alpha^2 - 2\beta s - s^2)^{1/2} ds = -\frac{1}{2}\alpha\beta$$

$$+ \frac{1}{2}(\alpha^2 + \beta^2)\{(\pi/2) - \tan^{-1}(\beta/\alpha)\}$$

$$\text{(A.4.10)}$$

### Details of integration of A.4.10

To perform the last integration, the integrand is modified as follows, with the substitutions given below.

$$\alpha^2 - 2\beta s - s^2 = (\alpha^2 + \beta^2) - (s + \beta)^2 \qquad \text{(A.4.11)}$$

$$= (\alpha^2 + \beta^2)\left[1 - \frac{(s + \beta)^2}{(\alpha^2 + \beta^2)}\right] = c^2(1 - y^2) \qquad \text{(A.4.12)}$$

Using substitutions:

$$c^2 = \alpha^2 + \beta^2 \qquad \text{(A.4.13)}$$

$$y = \frac{s + \beta}{\sqrt{\alpha^2 + \beta^2}} \qquad \text{(A.4.14)}$$

$$ds = \sqrt{\alpha^2 + \beta^2}\, dy \qquad \text{(A.4.15)}$$

Integration with respect to $s$ is converted to one with $y$ to $y$ is done as follows, where $s_1$ is the positive root of the quadratic equation, $s^2 + 2\beta s - \alpha^2 = 0$.

$$\int_0^{s_1} ds \rightarrow \int_{y_1}^{y_2} dy \qquad \text{(A.4.16)}$$

$$y_1 = \frac{\beta}{\sqrt{\alpha^2 + \beta^2}} \qquad \text{(A.4.17)}$$

$$y_2 = \frac{s_1 + \beta}{\sqrt{\alpha^2 + \beta^2}} = 1 \qquad (A.4.18)$$

where

$$(s^2 + 2\beta s - \alpha^2 = 0, \ s_{1,2} = -\beta \pm \sqrt{\alpha^2 + \beta^2}) \ (A.4.19)$$

Now, the original integral (A.4.10) is transformed as follows.

$$C \int_{y_1}^{y_2} \sqrt{1 - y^2} \, ds = (\alpha^2 + \beta^2) \times \int_{y_1}^{y_2} \sqrt{1 - y^2} \, dy \qquad (A.4.20)$$

This integration can be done by first substituting $y = \sin\theta$.

$$\int \sqrt{1 - y^2} \, dy = \int \sqrt{1 - \sin^2\theta} \, dy = \int \cos\theta \cdot \cos\theta \, d\theta \quad (A.4.21)$$

$$= \int \cos^2\theta \, d\theta = \int_{\theta_1}^{\theta_2} \frac{1 + \cos 2\theta}{2} \, d\theta$$

$$= \frac{1}{2}\theta \Big|_{\theta_1}^{\theta_2} + \frac{1}{4} \sin 2\theta \Big|_{\theta_1}^{\theta_2} \qquad (A.4.22)$$

$$= \frac{1}{2}\left[ \frac{\pi}{2} - \sin^{-1} \frac{\beta}{\sqrt{\alpha^2 + \beta^2}} \right]$$

$$+ \frac{1}{4}\left[ \sin\pi - \sin\left\{ 2\left( \sin^{-1} \frac{\beta}{\sqrt{\alpha^2 + \beta^2}} \right) \right\} \right]$$

$$= \frac{\pi}{4} - \frac{1}{2}\sin^{-1} \frac{\beta}{\sqrt{\alpha^2 + \beta^2}} - \frac{1}{4} \frac{2\alpha\beta}{(\alpha^2 + \beta^2)}$$

$$\qquad (A.4.23)$$

The identity of $\sin^{-1}\left( \dfrac{\beta}{\sqrt{\alpha^2 + \beta^2}} \right)$ and $\tan^{-1}\left( \dfrac{\beta}{\alpha} \right)$ is proved as follows.

$$\sin^{-1}\frac{\beta}{\sqrt{\alpha^2 + \beta^2}} = q \tag{A.4.24}$$

$$\sin q = \frac{\beta}{\sqrt{\alpha^2 + \beta^2}} \tag{A.4.25}$$

$$\cos q = \frac{\alpha}{\sqrt{\alpha^2 + \beta^2}} \tag{A.4.26}$$

$$\frac{\sin q}{\cos q} = \tan q = \frac{\beta}{\alpha} \tag{A.4.27}$$

$$q = \tan^{-1}\frac{\beta}{\alpha} \tag{A.4.28}$$

The last term in the result of integration can be converted to a much simpler form of $-\beta/\alpha$

$$\frac{1}{4}\sin 2\sin^{-1}\frac{\beta}{\sqrt{\alpha^2 + \beta^2}} \tag{A.4.29}$$

$$\sin^{-1}a = \theta \tag{A.4.30}$$

$$a = \sin\theta \tag{A.4.31}$$

$$\sin 2\theta = 2\sin\theta\cos\theta = 2\sin\theta(\sqrt{1 - \sin^2\theta}) \tag{A.4.32}$$

$$= 2a\sqrt{1 - a^2} \tag{A.4.33}$$

$$= 2\frac{\beta}{\sqrt{\alpha^2 + \beta^2}} \cdot \left(\frac{\alpha^2 + \beta^2 - \beta^2}{\alpha^2 + \beta^2}\right)^{1/2} = \frac{2\alpha\beta}{\alpha^2 + \beta^2} \tag{A.4.34}$$

By multiplying the above result by $(\alpha^2 + \beta^2)$, we obtain the result of the original integration with respect to $s$ as,

$$\int_0^{s_1} (\alpha^2 - 2\beta s - s^2)^{1/2} ds = -\frac{1}{2}\alpha\beta + \frac{1}{2}(\alpha^2 + \beta^2)$$
$$\left[\frac{\pi}{2} - \tan^{-1}\left(\frac{\beta}{\alpha}\right)\right] \quad (A.4.35)$$

which is identical to A.4.10.

We then integrate the above equation with respect to $\phi$ from 0 to $2\pi$. The term $\beta\alpha$ and $\tan^{-1}(\beta(\phi)/\alpha) = -\tan^{-1}(\beta(\phi + \pi)/\alpha)$ vanish when integrated with respect to $\phi$ between the limits 0 and $2\pi$ ($\int_0^{2\pi} \cos\phi d\phi = 0$), and then we get,

$$\bar{u}_z(r) = \frac{1-\nu^2}{\pi Y} \frac{p_0}{a} \int_0^{2\pi} \frac{\pi}{4}(\alpha^2 - r^2 + r^2 \cos^2\phi) d\phi \quad (A.4.36)$$

$$= \frac{1-\nu^2}{Y} \frac{\pi p_0}{4a}(2a^2 - r^2) \quad (A.4.37)$$

## A.4.1.3 Contact problem of two spheres

The geometry of two spheres that are in contact with each other is given in Figure A.4.2, where the origin of the coordinate coincides with the point of the first contact of the two spheres. $x$ and $y$ axes are the common tangent to at the contact point and $z$ axis is normal to this plane. The profiles of each surface in the region close to the origin is approximated by the expression of the form,

$$z_1 = A_1 x^2 + B_1 y^2 + C_1 xy + \cdots \quad (A.4.38)$$

By choosing the orientation of $x$ and $y$ axes, $x_1$ and $x_2$, so that the term $xy$ in the above equation vanishes,

$$z_1 = \frac{1}{2R_1}(x_1^2 + y_1^2) \quad (A.4.39)$$

where $R_1$ is the radius of sphere 1. Similarly for sphere 2, we have the following result.

$$z_2 = -\frac{1}{2R_2}(x_2^2 + y_2^2) \quad (A.4.40)$$

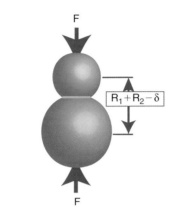

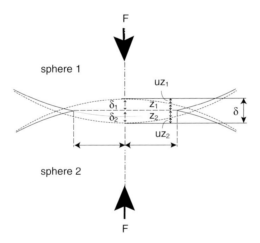

**Figure A.4.2** Two spheres in contact with each other under a compressive force, which flattens the two spheres at the contact site but without lateral extensions. Reproduced from Figure 4.2 in Ref. [1] with permission.

where $R_1$ and $R_2$ are the radius of spheres 1 and 2, respectively.

$$\bar{u}_{z1} + \bar{u}_{z2} + h = \delta_1 + \delta_2 \qquad (A.4.41)$$

The separation of the two surfaces at $(x, y)$ is then given by $h = z_1 - z_2$. By choosing a common set of axes $x$ and $y$, we have $h = Ax^2 + By^2 + Cxy$.

By choosing suitable set of axes,

$$h = Ax^2 + By^2 = \left(\frac{1}{2R_1} + \frac{1}{2R_2}\right)x^2 + \left(\frac{1}{2R_1} + \frac{1}{2R_2}\right)y^2$$

$$= \frac{1}{2R}(x^2 + y^2), \tag{A.4.42}$$

where, $\quad \dfrac{1}{R} = \dfrac{1}{R_1} + \dfrac{1}{R_2}$ \hfill (A.4.43)

$$A + B = \frac{1}{R_1} + \frac{1}{R_2} \tag{A.4.44}$$

$$\bar{u}_{z1} + \bar{u}_{z2} + h = \delta_1 + \delta_2 \tag{A.4.45}$$

Since $h = Ax^2 + By^2$,

$$\bar{u}_{z1} + \bar{u}_{z2} = \delta - Ax^2 - By^2 \tag{A.4.46}$$

From symmetry consideration, we set $A = B = 1/2R$ and by remembering that $x^2 + y^2 = r^2$, we have,

$$\bar{u}_{z1} + \bar{u}_{z2} = \delta - \frac{1}{2R^2}r^2 \quad \text{where} \quad \frac{1}{R} = \frac{1}{R_1} + \frac{1}{R_2} \tag{A.4.47}$$

The pressure distribution $p = p_0\{1 - (r/a)^2\}^{1/2}$ gives normal displacements of the following form as shown in the previous section.

$$\bar{u}_{z1} = \frac{1 - \nu_1^2}{Y_1} \frac{\pi p_0}{4a}(2a^2 - r^2), \ (r \le a) \tag{A.4.48}$$

The pressure acting on the second body is equal to that on the first, so that,

$$\frac{1}{Y^*} = \frac{1 - \nu_1^2}{Y_1} + \frac{1 - \nu_2^2}{Y_2} \tag{A.4.49}$$

Substituting $\bar{u}_{z1}$ and $\bar{u}_{z2}$ in Eq. (A.4.48), we get,

$$\frac{\pi p_0}{4aY^*}(2a^2 - r^2) = \delta - (1/2R)r^2 \qquad \text{(A.4.50)}$$

which, when $r = a$, give the radius of contact area as,

$$a = \frac{\pi p_0 R}{2Y^*} \qquad \text{(A.4.51)}$$

The total pressure, $P$,

$$P = \int_0^a p(r)2\pi r \mathrm{d}r = \frac{2}{3}p_0\pi a^2 \qquad \text{(A.4.52)}$$

The total pressure, $P$, can be replaced by the total applied force, $F$.

$$a = \left(\frac{3FR}{4Y^*}\right)^{1/3} \qquad \text{(A.4.53)}$$

$$\delta = \frac{a^2}{R} = \left(\frac{9F^2}{16RY^{*2}}\right)^{1/3} \qquad \text{(A.4.54)}$$

$$p_0 = \frac{3F}{2\pi a^2} = \left(\frac{6FY^{*2}}{\pi^3 R^2}\right)^{1/3} \qquad \text{(A.4.55)}$$

Thus, we obtain,

$$F = \frac{4}{3}\sqrt{R}Y^*\delta^{3/2} \qquad \text{(A.4.56)}$$

This equation is widely used for the analysis of indentation experiments and gives the value of Young's modulus of the sample under the assumption that the sample material is large, homogeneous, and isotropic.

## Bibliography

[1] Johnson, K. L. (1985), 'Contact Mechanics', p. 56, Cambridge University Press.

# INDEX